Dynamics in animal nutrition

Dynamics
in
animal nutrition

edited by:
Jannes Doppenberg
Piet van der Aar

ISBN: 978-90-8686-149-1
e-ISBN: 978-90-8686-706-6
DOI: 10.3921/978-90-8686-706-6

First published, 2010

Table of contents

Preface

In this book you will find the proceedings of the conference 'Dynamics in Animal Nutrition'. With this conference we celebrated the 75th anniversary of Schothorst Feed Research.

During the last 75 years animal production has changed dramatically. The technical results and production performance have increased considerably, the production efficiency has improved and lastly animal health and welfare are currently better than ever before. These achievements have been accomplished largely through increased understanding and knowledge of animal nutrition and science.

Schothorst Feed Research was established by farmer organisations in the Netherlands wanting to be assured of the nutritional quality of compound feeds. The production of manufactured feed just started to accelerate in that period. They insisted on an independent feed control from their suppliers. Initially this quality control was not limited to analysing the feed quality and nutrient composition but also included checking the prescribed feed composition and usage of feedstuffs. Consider that at this time period some vitamins and amino acids still had to be discovered. In order to be able to determine feedstuff restrictions and nutrient requirements as well as performing feeding experiments an experimental station for animal nutrition was established. Reduction of production cost and improving technical performances are no longer the sole drivers for animal production. Today, after 75 years, quality control and conducting nutritional research are still the main focus and core business of our organisation. We remain to be an independent organisation but are now more than ever internationally active.

The consumer's perspective has influenced and chanced the conditions for farming and animal production tremendously through regulation on a national and EU level. Today, society, consumers and food processors all have a strong interest in livestock production methods, quality of the animal products, product safety, product variety and assortment, housing conditions and animal health. This is driven by human health concerns and awareness of animal welfare. The feed industry is an integral part of the feed-food chain and can support the farmer in all the aspects of trying to meet these requirements by providing the (right) solutions via the feed and nutrition. Especially in Europe, the last decennium societal perceptions have affected the feed industry, forcing it to take measures to meet these demands. Examples are the ban on antibiotics, the use of GMO feedstuffs, the ban on meat and bone meal in animal feed, the discussion on the sustainability of the production of feedstuffs in ecologically important regions i.e. soybeans and palm oil, and the growing discussion about the competition between feed, food and fuel.

The public awareness on feed and animal production creates challenges for the feed industry, but also opportunities, especially for innovative organisations.

On top of these dynamics, the dominant global trends in the feed industry from our perspective for the next decennium will be:

1. Due to the increase in personal wealth and the world population the global consumption of animal products will increase further requiring more feedstuffs for animal production.
2. The scarcity of land suitable for the production of food, feed and fuel will result in higher prices for animal feedstuffs, and the need for knowledge of the use of non-traditional feedstuffs for the feeding of animals will increase.
3. Low production costs and increased management skills, both for the farmer and the feed supplier, will remain the most important factors determining the profitability and economical sustainability of individual producers.

Schothorst Feed Research desires to be the preferred partner to the feed-food chain assisting in providing the necessary expertise to meet these challenges for the future. For 75 years our know-how of feed formulation, feedstuff evaluation, the nutritional requirements for animal production, the effect of feed processing and the interaction between nutrition and gastro-intestinal health has been the fundamental base for providing adding added value to the feed industry. Our expertise, combined with the erudite knowledge and expertise of our global network of scientists in animal nutrition, is available for the feed industry now and in the future.

We are excited that so many leading renowned scientists from around the world are willing to share their knowledge with us during this conference. This together with all the participants from the feed and allied industry makes it a true international feed conference. The knowledge shared during this conference will be used in the consultancy with our ever increasing group of customers of Schothorst Feed Research.

We hope that this conference will give new insights, knowledge and perspectives in order to make strategic decisions for your organisation and cope with the challenges of this ever dynamic world of animal nutrition. Have a great and productive meeting.

Dr. Ir. P.J. van der Aar
Schothorst Feed Research B.V.

Understanding nutritional immunomodulation: Th1 versus Th2

Bruno M. Goddeeris
Division of Gene Technology, Department of Biosystems, K.U.Leuven, Kasteelpark Arenberg 30, 3001 Leuven, Belgium;
bruno.goddeeris@biw.kuleuven.be
Laboratory of Immunology, Faculty of Veterinary Medicine, UGent, Salisburylaan 133, 9820 Merelbeke, Belgium; bruno.goddeeris@ugent.be

Abstract

In order to modulate or influence the immune response or responsiveness of an animal by nutrition it is important to understand the basic immune effector mechanisms and factors influencing these. A fundamental knowledge on the intricate communication of the immunocompetent cells, their ontogeny and their induction sites and signals is required for understanding dietary modulation and/or enhancement of immune responsiveness. Moreover, a good understanding of the intimate relationship between the resident bacterial flora/pathogens and the host intestinal tract is quite important. In general we can distinguish two main categories in this search, a search for food components which have a direct effect on the pathogen or gut flora and secondly those which have an indirect effect on the gut pathogen or flora by targeting the intestinal and/or immune cells of the host. Some compounds (if not all) act even in both ways. The first category of compounds includes medium-chain fatty acids killing pathogens, sugars or lectins inhibiting colonization of the gut by competition or blocking, and competition in colonization by changes in the resident flora. The second group of modulators targets the innate immune system of the host by interacting with host receptors. To understand this innate immune alertness of the mucosal tract, it is imperative to focus on the acute phase of the inflammatory response. A lot of food components such as vitamins A and D, $\omega 6/\omega 3$ fatty acid ratios, oligosaccharides and others can indeed stimulate/modulate innate immunity at the mucosal site and protect against intestinal invaders. However, their mechanisms of action are not always fully understood and are now under thorough investigation. More recently, new groups of receptors, including the Toll-like and NOD receptors, have been discovered, which specifically recognize certain molecular patterns common in the microbial world, and referred to as pathogen-associated molecular patterns (PAMPs). These patterns are characteristic for lipopolysaccharides, lipoproteins, peptidoglycans, glycolipids and other molecules of microbial/yeast/fungal origin. They are present in abundance in the gut microbial flora and can be exploited in animal feeds to induce an immune alertness for mucosal protection against pathogens. Indeed, these PAMPS interact with pathogen recognition receptors (PRR) of the gut and, depending on the type of activated receptor, trigger warning signals that can modulate the acquired immune response towards inflammatory (T helper 1) or less inflammatory immune responses (T helper 2). In conclusion, this new scientific knowledge on communication between host and microbes in the intestinal world can yield innovative approaches to modulate immune responses towards less- or more inflammatory immune responses with protective and/or productive outcome. But does not everything has its price? Does immune alertness not compromise for energy at the expense of production?

1. Introduction

'Functional food/feed' is not experiencing the breakthrough that was expected some 10 years ago. The kind of mucosal immune status that is needed for optimal production protection or performance is not known. It is true that these goals are quite often contradictory goals: indeed an alerted immune status, even at mucosal surfaces, is often not beneficial for production or performance. Induction and maintenance (continuous alertness) of protective immune mechanisms will always be at the cost of energy which will consequently be lost for productive or other purposes. Nonetheless, a certain degree of mucosal immune alertness is needed to protect against chronic and acute infectious diseases that can cause major losses in production. Conversely overacting immune responses can be immunopathological and detrimental for the host. Therefore we will discuss factors that can direct immune responses towards the mucosae, can improve a general mucosal alertness or can reduce excess inflammation. A fundamental knowledge on the different nature of the immunocompetent cells, their intricate communication, their induction sites and communication signals is required for understanding the immune orchestra and its music and for dietary modulation of the tone of immune music.

2. The immune orchestra

An immune response is like *music*, produced by an *orchestra* (the immunocompetent cells) under the guidance of a *conductor* (the professional antigen-presenting cell such as the interdigitating dendritic cell, IDC) who interprets the *score* (antigen) (Figure 1). The immune

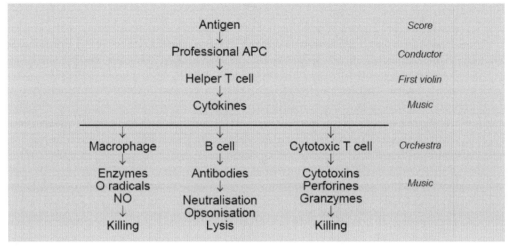

Figure 1. The immune orchestra, players and music.

responses can be divided in innate immunity, responsible for the first line of defense and induction of immune responses, and acquired immunity, responsible for antigen-specific memory. Innate immunity is mediated by the 'so-called' antigen a-specific immune cells such as the IDC (the conductor), macrophages and neutrophils (microphages). However, their important immune role is not restricted to a first-line defense but lays also in the triggering of alarm signals (*interpretation of the score and direction of the music*) which are crucial for initiating and directing subsequent antigen a-specific and specific defense reactions (the music). Moreover, recent data have shown that these 'so-called' a-specific cells, as well as other cells, are not as a-specific: indeed they can recognise and differentiate specific pathogen-associated molecular patterns (PAMPs) with pathogen recognition receptors (PRRs) (see next chapter). Depending on the PAMPs recognized, the IDC, as well as other PAMP activated cells, can direct the immune response into opposite ways with important implications on the inflammatory response, production, protection and performance.

The antigen specific cells (*members of the orchestra*) which have an immunological memory and secrete immuno-active molecules (*the music*), are divided in two main populations, the B and T cells (Figure 1). Their antigen specificity is mediated by cell surface receptors (*their instruments*), the B cell receptor (membrane-bound immunoglobulin, BCR) and the T cell receptor (TCR), respectively. B cells recognize with their BCR directly a specific epitope of an unprocessed antigen. T cells on the other hand, recognize with their TCR, epitopes of a processed antigen on the surface of an antigen-presenting cell (*conductor*). Indeed, the antigen is processed intracellularly (*interpretation of the score*) by an antigen-presenting cell (APC) into short peptides: linear epitopes of 8 to 20 amino acids long that associate intracellularly with major histocompatibility complex (MHC) molecules. These MHC molecules will, on subsequent expression on the cell membrane of the antigen-presenting cell, present the epitope to the T cell. The T cell population can be subdivided in $CD4^+$ and $CD8^+$ T cells. $CD4^+$ T cells recognize their epitope in association with MHC class II molecules while $CD8^+$ T cells recognize their epitope in association with MHC class I molecules. MHC class II molecules present epitopes from exogenous antigens, i.e. phagosomal antigens after uptake and processing by the APC (IDC, macrophage or B cell), while MHC class I molecules present epitopes from endogenous antigens, i.e. proteasomal antigens which have been transcribed inside the antigen-presenting cell and processed after ubiquitinilation in proteasomes, as is the case for virus-infected cells. This difference has very important consequences as $CD4^+$ T cells (or T helper cells, Th) compensate their antigen-presenting target cells with beneficial cytokines, while $CD8^+$ T cells kill their infected target cells with cytotoxic factors and signal transductions. Based on their cytokine-producing profiles, CD4+ T cells can be subdivided in different subtypes: the inflammatory Th cell or Th1 is characterized by the secretion of interleukin-2 (IL-2), interferon-γ (IFN-γ) and IL-12 while the antibody-promoting Th cell or Th2 is characterized by IL-4, IL-5, IL-10 and IL-13. Another Th cell has been categorized as Th3 and is responsible for the production of suppressive cytokines such as transforming growth factor β (TGF-β). The kind of $CD4^+$ T cell that will be induced, is determined by the APC (IDC *the conductor*), and the kind of cytokine(s) produced by the CD4+ T cell (*first violin and other solists*) will have important modulatory implications on the target cells and subsequently on the type of resultant immune responses (*music*), i.e. inflammatory or antibody response.

3. The induction of an immune response in secondary lymphoid organs

The induction of an antigen-specific immune response happens in the secondary lymphoid organs such as the tissue-draining lymph nodes and the blood-filtering spleen. It is here that IDC (the conductor) are key players in the activation of naïve T cells and the modulation of the response. Depending on their state and type of activation, they will induce naïve T cells into different effector T cells such as Th1, Th2 or regulatory T cells (tolerance).

These effector T cells will then activate specifically those B cells and macrophages presenting the same epitopes as did their inducing IDC before. One could say that B cells and macrophages present antigen to T cells for their own benefit because they need cytokines and other stimulatory molecules from the T cell to be activated for production of antibodies and enzymes, respectively. Moreover, the ratios between the different types of cytokines produced by the different T helper cells will determine which antibody class (IgG, IgA or IgE) or immune response (inflammatory or less inflammatory) will be produced.

The state and type of activation of IDC will not only depend on the type of antigen to be presented but also on the microenvironment where it happens. With microenvironment we mean all type of signals which can be perceived by the IDC from their surroundings. This means that different environments like mucosal versus systemic can have an influence on the type of activation of the IDC and consequently on the kind of T cell responses which will issue. Moreover, mucosal IDC are in contact with the gut lumen through their dendrites between the enterocytes and can detect with their receptors signals from the intestinal flora and pathogens such as PAMPs. It is in this light that the next chapters will discuss some of these receptors and the microenvironment (acute phase response) which can alter the stimulatory behaviour of IDC.

4. Immunological defence of the gut and the gut-associated lymphoid tissue (GALT)

The immune defence system of the gut consists of lymphoid tissues and cells distributed along the gastrointestinal tract (reviewed in Brandtzaeg and Pabst, 2004). The lymphoid tissue localized along the gastrointestinal tract constitutes quantitatively the major part of the immune system of the whole body. This extremely developed gastrointestinal immune system reflects the importance of the mucosal immune defence system against the continuous attack of antigens and pathogens. Moreover, the major development of this local immune tissue seems to be induced by the continuous contact of the gastro-intestinal mucosa with the gastrointestinal flora, as well with pathogens, as evidenced by the atrophic mucosal immune system in axenic (germ-free) animals (Smith et al., 2007).

Some important features characterize the mucosal immune system:
- it possesses mucosa-associated lymphoid tissue (MALT or GALT for the gut-associated lymphoid tissue) as well as local and regional draining lymph nodes where the induction of immune responses is established, such as the Peyer's patches (PP) and the mesenterial lymph nodes, respectively;
- certain subpopulations of lymphoid cells predominate at the mucosal surfaces;
- there is a specific recirculation of mucosal lymphocytes towards mucosae, known as mucosal homing;

- the predominant mucosal immunoglobulin is dimeric IgA which is secreted at the mucosal surface.

All these elements of the mucosal immune system are working together to generate immune responses that protect the host against mucosal invaders but also render the host tolerant against ubiquitous dietary antigens and the beneficial microbial flora of the mucosae. The elucidation of the mechanisms that determine immunotolerance or immunoinduction forms today a major topic of study where the type of activation state of the antigen-presenting IDC will determine whether T cells will or will not differentiate in T regulatory cells playing a role in immunotolerance.

The gastrointestinal immune system consists of lymphoid cells in organised sites like PP and mesenterial lymph nodes, and lymphocytes spread over the stromal tissues in the lamina propria and the epithelium (the intra-epithelial lymphocytes, IEL) of the intestine. There exists evidence that these organized structures are the places (or one of the places) where antigen enters the mucosal immune system to initiate subsequently the immune reactions: therefore these places are quite often called the inductive lymphoid sites of the mucosal immune system. Lymphocytes located in the lamina propria and between the epithelial cells of the intestinal tract are attributed with effector functions, such as antibody production, cytokine production and cytotoxicity, and these places are therefore referred to as the effector sites of the intestinal tract. These inductive and effector sites are interconnected by selective migration of lymphocytes (homing) whereby cells that have been activated in the inductive sites migrate specifically to effector sites of the intestinal tract. This interconnection assures that mucosal responses are primarily directed and localised against antigens that have been recognized at mucosal surfaces.

5. Pathogen recognition receptors (PRR) and the pathogen-associated molecular patterns (PAMP)

As mentioned above, new families of receptors, namely PRRs, have been discovered which recognize specific patterns on molecules common to the microbial world, referred to as PAMPs and play an important role in the innate immunity. Indeed, to detect microorganisms, animals (but also plants) have evolved 4 large gene families of PRRs: two 'leucine-rich repeats-containing' families, namely the membrane-bound TLR (Toll-like receptors) family and the cytosolic NLR (nucleotide-binding domain, leucine-rich repeats containing) protein family (Parker *et al.*, 2007; Wilmanski *et al.*, 2008), the cytosolic RLR (retinoic acid-inducible gene 1-like receptors) family and the CLR (C-type lectin receptors) family

(Palsson-McDermott and O'Neill, 2007). The PAMPs are characteristic for lipopolysaccharides, lipoproteins, peptidoglycans, glycolipids, RNA and DNA and other molecules of bacterial, viral, yeast or fungal origin (Table 1). Upon recognition of their PAMP ligands, these PRR produce or induce the production of immune effector molecules, e.g. interferons and pro-inflammatory cytokines, which can modulate immune responses.

In our experiments in poultry and pigs, we demonstrated enhanced antigen-specific antibody responses upon co-injection of antigen with CpG oligodinucleotides which stimulate the TLR9. Different PAMPs are present on the intestinal flora and pathogens and could thus be exploited in animal feeds for modulation of mucosal immune responses and protection against pathogens. DC which are activated by these PAMPS through a TLR or NLR can depending on the type of activated receptor, induce and/or modulate the ensuing response towards inflammatory (Th1) or less inflammatory immune responses (Th2)

Table 1. List of 'pathogen recognition receptors' (PRR) families.

Family	Receptor	Ligand
TLR family (sensing bacteria, viruses, protozoa and fungi)		
Nucleic acid ligands (endosomes)	TLR3	dsRNA
	TLR7	ssRNA
	TLR8	ssRNA
	TLR9	CpG DNA
Lipid ligands (cell membrane)	TLR1/2	diacyl-lipopeptides
	TLR2/10	diacyl-lipopeptides
	TLR6/2	triacyl-lipopeptides
	TLR4/CD14/MD2	triacyl-lipopeptides (LPS)
Protein ligands (cell membrane)	TLR5	flagellin
	IL-1R	IL-1, IL-18
NLR family (sensing bacteria)		
Sugar ligands (cytosol)	NOD-1	peptidoglycans (DAP)
	NOD-2	peptidoglucans (MDP)
	Nalp1	peptidoglycan (MDP)
Protein ligands (cytosol)	Ipaf	flagellin
	Naip	flagellin
	Nalp1	anthrax lethal toxin
Nucleic acid ligands (cytosol)	Nalp3	RNA
RLR family (sensing viruses)		
Nucleic acid ligands (cytosol)	RIG-I	dsRNA and ssRNA
	MDA 5	dsRNA
CLR family (sensing fungi)		
Sugar ligands (cell membrane)	Dectin-1	β-glucan (zymosan)

(Germain 2004). Interesting to note for the next chapter that TLR2 also increases cyclooxygenase 2 expression, responsible for production of eicosanoids (see next chapter).

6. The acute phase response and the arachidonic acid cycle

Inflammation is the answer of tissue to irritation, injury and infection, and is quite often needed for the induction of strong immune responses. It consists of a complex cascade of non specific events, known as the acute-phase response, which confer early protection by limiting tissue injury to the place of infection or destruction (Figure 2). One important function of this reaction is to recruit more phagocytic cells to the site of injury. Moreover, it initiates the specific immune response against the invader. The localized reaction is induced by clotting factors and pro-inflammatory cytokines released by the activated resident sentinels, the tissue macrophages. The combined actions of their pro-inflammatory cytokines IL-1, IL-6, IL-12 and TNF-α and the release of chemokines by activated macrophages and activated structural tissue cells (keratinocytes, fibroblasts, endothelial and epithelial cells) are responsible for changes in the surrounding capillaries, inducing an influx of neutrophils, monocytes and effector lymphocytes into the site of inflammation. Indeed, there is an increase of inflammatory adhesion molecules on the endothelial cells, which trap the circulating leukocytes

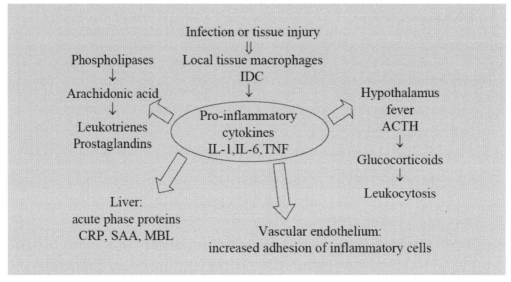

Figure 2. Acute phase response (CRP C-reactive protein, SAA serum amyloid A, MBL mannan-binding lectin, ACTH adrenocorticotropic hormone, IL interleukin, TNF tumor necrosis factor).

on the endothelium with subsequent diapedesis and migration of these cells towards the place of tissue injury for fulfillment of their duties.

However, activated phagocytes release also other proteins with potent local effects, such as toxic radicals, peroxides, nitric oxide, plasminogen activator and enzymes. Indeed the enzyme phospholipase A_2 cleaves the fatty acid arachidonic acid (C20:4n6) from the glycerol backbone of membrane-bound phospholipids. The liberated arachidonic acid can then undergo controlled oxidative metabolism to form a variety of eicosanoids (C20, eicosa=twenty) with different physiological and immune effects. The cyclooxygenase pathway yields prostaglandins (PG), prostacyclin and thromboxanes while the lipoxygenase pathway produces the leukotrienes (LT) and lipoxins (Figure 3). The eicosanoids act as autocrine/paracrine regulators, since most of their biological effects are limited to the site of biosynthesis. LTB_4 stimulates chemotaxis of neutrophils and an increased expression of their C3b receptors while LTC_4, LTD_4 and LTE_4 are collectively known as the slow reactive substance of anaphylaxis (SRS-A), being more than 1000 times more potent than histamine in smooth muscle contraction and vasodilatation. The latter is important in view of the increased blood flow in inflamed tissues inducing an even higher influx of leukocytes into the inflammation site. After exerting their effect, eicosanoids are rapidly catabolised to inactive compounds in the liver and lungs.

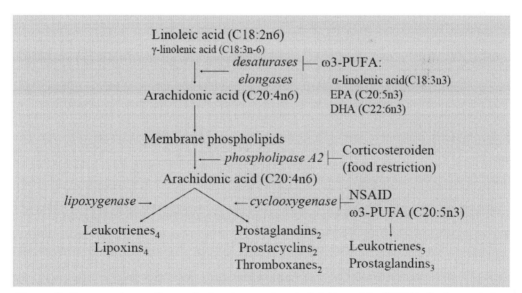

Figure 3. Suppressive actions of ω3-poly-unsaturated fatty acids (PUFA), food restriction, non-steroidal anti-inflammatory drugs (NSAID) on the arachidonic acid cycle and its inflammatory products. Left part: the arachidonic acid cycle; right part: the inhibiting or suppressive molecules (EPA eicosapentaenoic acid; DHA docosahexaenoic acid).

7. Nutritional modulation of immune responses

Primary induction of the acquired immune responses occurs in the secondary lymphoid organs which drain the site of infection. These structures are strategically placed on the lymph (lymph nodes) and blood circulation (spleen) to filter the antigen and antigen-presenting cells for optimal encounter with the T and B cells. By specific adhesion molecules, naive lymphocytes are directed to emigrate from the blood circulation into these secondary lymphoid organs in search for their specific antigen. For induction of a primary immune response and its subsequent direction, the crucial antigen-presenting cell is the IDC, which is present in all tissues of the host. Depending on the signals it receives from the environment, the IDC will secrete specific cytokines which determine in which type the Th cells will differentiate, i.e. Th1, Th2, Th3 or another type. The cytokine environment of IL-12, IL-18, IL-23, IL-27, IFN-γ and IFN-α will induce the differentiation of Th cells into Th1 while the cytokine environment of IL-4, IL-6, IL-10 and IL-13 into Th2.

The intestinal mucosal wall comes into contact with a diversity of antigens. Nutrients, intestinal bacteria and their products, and pathogens interact with the mucosal cells and influence their metabolic behaviour as well as that of the whole host. Moreover and perhaps not surprisingly, a majority of the immunocompetent cells are directly or indirectly in contact with the intestinal tract that modulates immune responses towards induction or tolerance. These immunocompetent cells are free or organised in lymphoid structures (Peyer's patches) within the intestinal wall, or in lymph nodes (mesenterial) draining the intestinal tract. A fundamental knowledge on the intricate communication of these immunocompetent cells and signals is required for understanding dietary modulation of immune responsiveness. In general we can distinguish two directions in the search for antibacterial mechanisms, first a search for food components which have a direct effect on the pathogen or gut flora and secondly a search for those which have an indirect effect on the gut pathogen or flora by targeting the intestinal and/or immune cells of the host (Figure 4). Some compounds might even act in both ways.

The first category of compounds includes medium-chain fatty acids (MCFA) killing directly pathogens, and sugars or lectins inhibiting colonization of the gut by competition or blocking or changes in the resident flora. The second group of modulators targets the innate immune system of the host by interacting with host receptors. To understand this innate immune alertness of the mucosal tract, it is imperative to focus on the acute phase of the inflammatory response as explained above. A lot of food components such as vitamins, $\omega 3/\omega 6$ fatty acid ratios, oligosaccharides and others can indeed stimulate/modulate

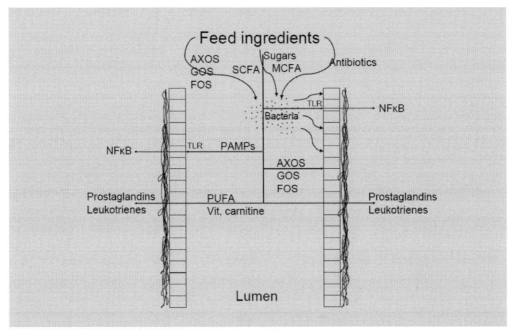

Figure 4. The effect of food ingredients and intestinal bacteria on immune modulators (AXOS arabinoxylan-oligosaccharides, GOS gluco- oligosaccharides, FOS fructo-oligosaccharides, SCFA short chain fatty acids, MCFA medium chain fatty acids, PUFA poly-unsaturated fatty acids, TLR Toll-like receptors, NFκB nuclear factor kappaB).

innate immunity (including the IDC) at the mucosal site and protect against intestinal invaders. However, their mechanisms of action are not always fully understood, but data indicate that they can modulate immune responses towards Th1 (inflammatory) responses or towards Th2 responses.

Vitamin A and D₃

After passing through the cell membrane, liposoluble vitamin D3 can bind to the cytoplasmic/nuclear $1,25D_3$-receptor (VDR) and will exert its regulation of gene expression in target cells by binding to DNA D_3-responsive elements. The binding of VDR to its D_3-responsive elements requires the presence of its ligand $1,25(OH)_2D_3$ and a companion protein belonging to the RXR group of retinoic acid (vitamin A) receptors (Figure 5). VDR and RXR respectively, are members of the same superfamily of ligand-activated transcription factors and act together as heterodimers in their function. This demonstrates that vitamin D and vitamin A are linked in their signaling mechanisms. Vitamin D and A (retinol) appear to play rather an opposing role in the

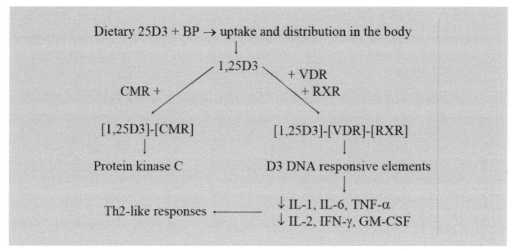

Figure 5. Influence of vitamin D3 on the Th1/Th2 balance.

control of lymphocyte migration towards mucosae or skin, but a similar role in directing immune responses towards Th2.

- DC of the skin are able to convert vitamin D3 (after exposure to sunlight) to its active form $1,25(OH)_2D_3$ which upon binding on its nuclear VDR results (together with IL-12) in the expression of the chemokine receptor CCR10. The latter allows the T cells to migrate towards the chemokine CCL27 produced by keratinocytes of the epidermis (Mebius, 2007).

- DC in the GALT express enzymes that convert vitamin A into retinoic acid. T cell as well as B cell activation in the presence of retinoic acid subsequently leads via the nuclear retinoic acid receptor to the induction of gut-homing molecules $\alpha4\beta7$ and the chemokine receptor CCR9, binding respectively to MAdCAM-1 and the chemokine CCL25 in the mucosae. Moreover, together with IL-5 and IL-6, retinoic acid promotes the production of IgA. Conversely, retinoic acid is also able to suppress the expression of dermis homing receptors (CCR4 and E- and P-selectin ligands) on T cells inhibiting consequently migration towards the skin (Mebius, 2007).

- Vitamin A inhibits secretion of Th1 cytokines (IFN-γ, GM-CSF, IL-2) but not Th2 cytokines (IL-4, IL-10), possibly through an inhibitory effect on protein kinase C activity. We observed also Th2-like effects in pigs upon intramuscular co-administration of vitamin D3 with antigen (Van der Stede et al., 2001, 2003) with better mucosal immune responses.

Thus all the hormonal/immunological effects of these liposoluble vitamins open possibilities for modulating immune responses towards Th2 and consequently less or anti-inflammatory immune responses.

Ratio of ω3/ω6 PUFA in feed

As we explained above, arachidonic acid (C20:4n6) is liberated during the acute-phase response from membrane phospholipids by the action of phospholipase A2 and is ultimately responsible for the generation of inflammatory eicosanoids (Figure 3). The essential polyunsaturated fatty acid (PUFA) linoleic acid (C18:2n6) is the nutritional source for the generation of arachidonic acid containing phospholipids. Indeed, linoleic acid is converted by desaturases and elongases to eicosatetraenoic acid (C20:4n6) which is incorporated in membrane phospholipids. However when significant amounts of α-linolenic acid (C18:3n3) are present in the feed, the concentration of ω-3 PUFA increases with consequences in the composition of the membrane phospholipids and the production of less inflammatory eicosanoids (PG_3 and LT_5 series). Nutritional intervention to modulate eicosanoid production is thus primarily mediated by competition between ω-6 and ω-3 PUFA for desaturases/elongases, or by ω-3 PUFA inhibition of cyclooxygenase. Thus, dietary ω-6 and ω-3 PUFA determine the type and production rate of inflammatory eicosanoid in leukocytes and accessory cells, and modulates the immune response towards inflammatory (PG_2 and LT_4) or less inflammatory (PG_3 and LT_5) eicosanoids, respectively. It appears that a decrease in the ω-6/ω-3 dietary PUFA ratio favours antibody responses over cell-mediated (inflammatory) responses.

Feed restriction

Feed restricting or fasting induces increased levels of glucocorticoids, glucagon, insulin and growth hormone and decreased levels of thyroid hormones, having as a consequence a glucose decrease and free fatty acid increase in the blood. But these increased corticosteroid levels have also an influence on the immune system. They are responsible for a detachment and redistribution of heterophils from the marginal pool (endothelium of the blood vessels) to the tissues. Moreover, glucocorticoids have an important impact on the transcription of cytokine and other genes that play an important role in immune responses. Indeed, the liposoluble glucocorticoids (just like $VitD_3$ and VitA) are able to pass the cell membrane into cytoplasm of leukocytes where they bind to their receptors which can subsequently migrate into the nucleus to associate with transcription-regulating sequences on DNA. The expression of as many as 1% of the genes might be regulated by glucocorticoids: they can either up-regulate or down-regulate the responsive genes. From an immunological viewpoint, this is quite important as several cytokine and immunomodulating genes are here the targets. Indeed, suppression of phospholipase A_2 by glucocorticoids reduces the eicosanoids with resultant reduction in inflammatory responses and a move towards Th2 responses (Figure 3). This effect is similar to an increase of ω-3 PUFA in feed.

L-carnitine

The mechanism(s) accounting for the positive effect of L-carnitine on antibody production are currently not clear. Also in poultry, we have seen that dietary L-carnitine had positive effects on antigen-specific antibody responses upon immunization (Mast *et al.*, 2000) but also on the acute phase response upon lipopolysaccharide injection (Buyse *et al.*, 2007). Restoration of the cellular L-carnitine content might enhance the lipid metabolism and improve the cellular energy balance. L-carnitine decreases the concentrations of cytokines, most notably TNF-α, IL-1β, IL-6 and TNF-α. These cytokines play a pivotal role in general energy homeostasis, but also in the modulation of antibody responses. They appear to down-regulate the inflammatory responses. Consequently, L-carnitine yields higher antibody titers and plausibly also prevents apoptotic cell death of B and T lymphocytes during the immune response. Moreover, L-carnitine has recently been shown to interact with the glycocorticoid receptor (Alesci *et al.*, 2004) activating the same genes responsible for directing the immune responses towards Th2 which decreases inflammatory responses and increases antibody responses.

β-glucans

Many studies in mammalians demonstrate that soluble and particulate β-glucans are immunological response modifiers and can be used in the therapy of experimental neoplasia, infectious diseases and immunosuppression. Their main direct target cells appear to be the monocyte/macrophage, neutrophil and natural killer cell. Glucan treatment of monocyte/macrophages induces the production of TNF-α, activation of IL-1, platelet-activating factor and phospholipase A_2 (Suram *et al.*, 2006) with formation of arachidonic acid and its metabolites, such as PGE_2 and LTB_4. The direct interaction of β-glucans with their target cell, the macrophage, is mediated by cell membrane receptors such as the lectin-binding domain of the α chain (CD11b) of the complement receptor 3 and the membrane-bound lectin dectine-1 belonging to the CLR family of PRRs, and probably also by TLRs. All these effects would indicate that these immune responses help in protection against intestinal infections but can (by excess) also lead to strong inflammation at the expense of production. Indeed, from our experiments in pigs, too high concentrations in the diet appear to be negative for production. However, it is very important to note here that we have observed different and sometimes opposing effects between β-glucans from different origin or structure; in other words not all β-glucans have similar effects or behave in a similar way (Stuyven *et al.*, 2009; Sonck *et al.*, 2009).

8. Conclusion

All this new scientific knowledge makes the interaction between nutrition and immunity more complex but can yield innovative approaches to modulate immune responses towards less-inflammatory immune responses with beneficial effects on health, production or physical performance.

References

Alesci, S., De Martino, M.U., Kino, T., Ilias, I., 2004. L-carnitine is a modulator of the glucocorticoid receptor alpha. Ann. N.Y. Acad. Sci. 1024: 147-152.

Brandtzaeg, P., Pabst, R., 2004. Let's go mucosal: communication on slippery ground. Trends Immunol 25: 570-577.

Buyse, J., Swennen, Q., Niewold, T., Klasing, K.C., Janssens, G.P.J., Baumgartner, M., Goddeeris, B.M., 2007. Dietary L-carnithine supplementation enhances the lipopolysaccharide-induced acute phase protein response in broiler chickens. Vet Immunol Immunopathol 118: 154-159.

Germain, R.N., 2004. An innately interesting decade of research in immunology. Nat. Med. 10: 1307-1320.

Mast, J., Buyse, J., Goddeeris, B.M., 2000. Dietary L-carnitine supplementation increases antigen-specific immunoglobulin (Ig) G in broiler chickens. Brit J Nutr 83: 161-166.

Mebius, R.E., 2007. Vitamins in control of lymphocyte migration. Nat Immunol 8: 229-230.

Palsson-McDermott, E.M., O'Neill, L.A.J., 2007. Building an immune system from nine domains. Biochem Soc Transact 35: 1437-1444.

Parker, L.C., Prince, L.R., Sabroe, I., 2007. Translational mini-review series on Toll-like receptors: networks regulated by Toll-like receptors mediate innate and adaptive immunity. Clin Exp Immunol 147: 199-207.

Smith, K., McKoy, K.D., Macpherson, A.J., 2007. Use of axenic animals in studying the adaptation of mammals to their commensal intestinal microbiota. Sem Immunol 19: 59-69.

Sonck, E., Stuyven, E., Goddeeris, B.M., Cox E., 2009. The effect of beta-glucans on porcine leukocytes. Vet Immunol Immunopathol 135: 199-207.

Stuyven E., Cox E., Vancaeneghem S., Arnouts S., Deprez P., Goddeeris B.M., 2009. Effect of beta-glucans on an ETEC infection in piglets. Vet Immunol Immunopathol 128: 60-66

Suram, S., Brown, G.D., Ghosh, M., Gordon, S., Loper, R., Taylor, P.R., Akira, S., Uematsu, S., Williams, D.L., Leslie, C.C., 2006. Regulation of cytosolic phospholipase A_2 activation and cyclooxygenase 2 expression in macrophages by the β-glucan receptor. J Biol Chem 281: 5506-5514.

Wilmanski, J.M., Petnicki-Ocwieja, T., Kobayashi, K.S., 2008. NLR proteins: integral members of innate immunity and mediators of inflammatory diseases. J Leuk Biol 83: 13-30.

Van der Stede, Y., Verdonck, F., Vancaeneghem, S., Cox, E., Goddeeris, B.M., 2002. CpG-oligodinucleotides as an effective adjuvant in pigs for intramuscular immunizations. Vet Immunol Immunopathol 86: 31-41.

Van der Stede, Y., Cox, E., Van den Broeck, W., Goddeeris, B.M., 2001. Enhanced induction of IgA in pigs by calcitriol after intramuscular immunisation. Vaccine 19: 1870-1878.

Basis and regulation of gut barrier function and epithelial cell protection:

applications to the weaned pig

J.P. Lallès
Institut National de la Recherche Agronomique (INRA), UMR1079, Systèmes d'Elevage, Nutrition Animale et Humaine (SENAH), 35590 Saint-Gilles, France; Jean-Paul.Lalles@rennes.inra.fr

Abstract

This review focuses on two important aspects of gut function: the epithelial barrier and its neuro-immune regulation, and epithelial cell protection by heat shock proteins (HSP). Basic aspects of gut homeostasis are summarized and their applied relevance to weaned pigs is provided. Major progress in the understanding of gut barrier function and epithelial cell protection and their regulation has been made recently. Tight junctions and HSP (especially HSP 25 and HSP 70) appear to have a major role in gut defence. Experimental evidence from rodent models indicates that these defence systems are highly dynamic, integrated and finely regulated at the epithelial monolayer. The first level of gut barrier regulation is a direct cross-talk between the host and the gut microbiota. The second regulatory level involves the nervous system, mucosal mast cells and various mediators acting on the epithelium. Gut barrier function and its regulation can be altered following stress or after disturbing the microbiota. These findings have very recently led to the concept that the microbiota are an integrate component of the brain-gut axis. Adverse events (e.g. stress, antibiotic therapy) occurring early in life may have negative consequences on gut barrier function in adulthood, emphasizing the importance of gut bacterial colonisation during the neonatal period in the development of an optimal gut physiology. Interestingly, the first available data suggest that long-term alterations in gut barrier function may be alleviated, at least partly through nutrition intervention. Studies are still scarce in the swine species and, therefore, merit further developments in the areas of gut barrier function, epithelial cell protection and their regulations. This in turn will help developing appropriate strategies for preventing gut disorders, including those observed after weaning. Long-term issues should also be considered.

1. Introduction

Intensive pig production has led to the development of so-called post-weaning disorders that favour gut infections and are responsible for large economical losses in the swine industry. The weaning transition is complex, associating psychological, social, nutritional and environmental stress (review by Pluske *et al.*, 1997). Early weaning of piglets is often accompanied by a severe growth check and diarrhoea. This process is considered as multifactorial, and anorexia and the resulting under-nutrition seen after weaning are major aetiological factors. Gastrointestinal disturbances include alterations in small intestinal architecture and enzyme activities. Recent data indicate transient increase in mucosal permeability, disturbed absorptive-secretory electrolyte balance and altered local inflammatory cytokine patterns (review by Lallès *et al.*, 2004). Early weaning (e.g. at 3-4 weeks of age) as in intensive production systems has probably increased the level of stress perceived by these immature pigs. Post-weaning gut disorders were kept under control using low levels of antibiotics in weaning diets for decades until their ban in the European Union in January 2006. Many nutritional alternatives have been evaluated to overcome these gut disorders but only some of them (e.g. glutamine, specific organic acids and spray-dried plasma proteins) proved to be satisfactory (reviews by Gallois *et al.*, 2009; Lallès *et al.*, 2009). This low effectiveness may be explained, at least partly by the incomplete understanding of the cellular, molecular and regulatory mechanisms underlying gut functional alterations observed after weaning.

The primary function of the gut is to digest food components and to absorb nutrients released by digestion. Its second function is to protect the organism against the entry of potentially harmful food constituents, pathogenic bacteria and viruses. This protection is brought about by complementary components (Table 1) and functions, including nutrient absorption, electrolyte and water secretion, epithelial cell proliferation and differentiation, and epithelial restitution after damage. The epithelial layer appears as a major component of the gut barrier function, involving both between- and within-cell protection systems, namely tight junctions (TJ) and intra-cellular heat shock proteins (HSP).

In this paper, two important aspects of gut function are reviewed: the epithelial barrier and its neuro-immune regulation on the one hand, and epithelial cell protection by HSP on the other hand. Examples of how nutrient absorption *per se* can also contribute actively to gut protection are provided. Basic aspects of gut homeostasis are summarized and their applied relevance to weaned pigs is shown. Finally, a focus is given on the neonatal period during which regulatory mechanisms of gut barrier function are set. Environmental factors, including stress

Table 1. Components of the gut barrier.

Mucus	secreted into the lumen
Defensins	
Immunoglobulin A (IgA)	
Digestive enzymes and peptides	
Epithelium	**epithelial barrier**
enterocytes/colonocytes	
tight junctions/proteins (e.g. occludin, claudins, zonula occludens)	
cell protection systems (e.g. HSP)	
Immune cell network (intra-epithelial lymphocytes, dendritic cells, T and B lymphocytes, mast cells, etc.)	lamina propria

and early antibiotic therapy are able to disturb both gut bacterial colonisation and barrier function.

2. Glucose absorption, barrier function and epithelial cell protection

Glucose absorption

Glucose is an essential fuel for the small intestine and the rest of body tissues and organs. Earlier research has shown that sodium-dependent glucose absorption by itself increases epithelial barrier permeability physiologically, thus enhancing glucose passive absorption greatly (60 to 90% of total glucose absorption; review by Pappenheimer, 1988). The underlying mechanism involved the contraction of the TJ actin-myosin ring. Recent data shed new light on the role of active glucose transport on intestinal epithelial barrier protection. The up-regulation of the sodium-dependent glucose cotransporter SGLT-1 in enterocyte apical membrane specifically attenuates lipopolysaccharide (LPS)-induced paracellular permeability (Yu *et al.*, 2005). The mechanisms involve the inhibition of the mitochondrial release of cytochrome c to the cytosol and the up-regulation of anti-apoptotic genes (Bcl-2 and Bcl-X) by high glucose concentrations. Interestingly, high glucose-induced SGLT-1 transporter activation mediates the protection of intestinal epithelial cells (IEC) to parasite infection (Yu *et al.*, 2008) and that of mice to lethal endotoxic shock (Palazzo *et al.*, 2008). Therefore, active intestinal absorption of nutrients must be borne in mind as participating directly in gut protection.

Gut barrier structure, function and regulation

Within gut barrier components, the epithelial layer plays a major dynamic role. Individual intestinal/colonic epithelial cells are connected together thanks to intercellular junctions, including (from cell apical to basal pole): TJ, adherens junctions, desmosomes and gap junctions (review by Ma and Anderson, 2006). Other cell types, including goblet cells, M cells, dendritic cells and intra-epithelial lymphocytes are present in the epithelial layer and display TJ with neighbour cells. Tight junctions are the most important structures of the gut barrier, making links with the cytoskeleton and actin filaments, and controlling the so-called para-cellular permeability. Many (> 40) different and highly conserved proteins have been identified in the TJ complex and their function elucidated (review by Harhaj and Antonetti, 2004). They include trans-membrane proteins involved in cell-cell adhesion and regulation of immune cell trans-migration. Claudins (20-25 kDa) form a large (> 20) family of trans-membrane proteins which contribute to the formation of TJ strands, confer cell-cell adhesion and determine selective size, charge and electric properties of the paracellular pathway. Another important trans-membrane protein, occludin (40 kDa) is localised in TJ strands and its expression is correlated with enhanced barrier properties. Occludin may be involved in the coordination of the numerous signalling pathways (e.g. protein kinase C, PKC) located at the TJ. Scaffolding proteins of the zonula occludens (ZO) family (ZO-1, ZO-2 and ZO-3) play a role in the organization of TJ complexes and may regulate epithelial cell growth and differentiation.

Gut epithelial TJ are permanently modulated by various extra-cellular physiological, pharmacological, bacterial and inflammatory molecules acting to activate intra-cellular signaling pathways (Ma and Anderson, 2006). Tight junction permeability may be regulated directly through changes in TJ proteins or indirectly via the cytoskeleton. One central mechanism in this regulation involves myosin light chain kinase (MLCK) which modulates the cytoskeleton-TJ protein interactions via actin-ZO-1 binding, ultimately increasing gut permeability. Many other regulatory pathways (e.g. Rho, PKC) may also cause a disruption of TJ resulting in an increase in epithelial permeability (Harhaj and Antonetti, 2004).

Gut barrier function is highly sensitive to acute and chronic stress (Gareau et al., 2008; Soderhölm and Perdue, 2006). Responses to stress involve mainly the production of corticotrophin-releasing factor (CRF) by the hypothalamus, its release and its action through central and peripheral CRF receptors (CRF-R1 and CRF-R2). CRF pathway activation leads to alterations of many aspects of GIT function, including epithelial permeability (Table 2).

Table 2. CRF-mediated alterations in GIT function and respective implications of CRF receptors 1 and 2 (adapted from Gareau et al., 2008).

CRF-R1	CRF-R2
Colonic transit (increased)	Gastric emptying (delayed)
Colonic sensitivity (increased)	Visceral pain (prevented)
Paracellular permeability (increased)	Transcellular permeability (increased)
Visceral pain (increased)	

Another central player of gut barrier dysfunction is the mucosal mast cell. It can be activated by nervous pathways directly or indirectly (via CRF) to release an array of chemical mediators responsible for increased epithelial permeability (Gareau *et al.*, 2008; Soderholm and Perdue, 2006) (Figure 1).

In terms of mediators, cytokines produced by various types of mucosal cells (e.g. macrophages, activated T cells, mast cells) are major regulators of intestinal epithelial permeability (reviews by McKay and Baird, 1999; Al-Sadi *et al.*, 2009). Inflammatory cytokines (e.g. IL-1α,

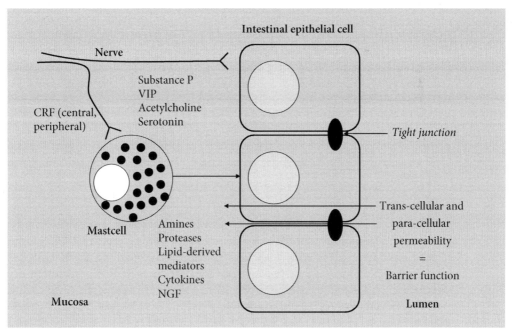

Figure 1. Neuroimmune control of intestinal epithelial barrier function (adapted from Gareau et al., 2008).

IL-4, IL-13, IFNγ, TNF-α) usually increase paracellular permeability and anti-inflammatory cytokines (e.g. IL-10, TGF-β1) decrease it (Table 3). However, exceptions and interactions among cytokines on epithelial barrier do exist.

Finally, stress activation of the hypothalamic-pituitary-adrenal axis and peripheral release of glucocorticoids is little involved in gut permeability alterations (Gareau et al., 2008).

Cell protection of the gut epithelium

Tremendous numbers of different proteins are synthesized permanently within cells. This requires a dynamic molecular chaperoning system for favouring native conformation and limiting aggregation of proteins in the normal unstressed situation. This system is also crucial for preventing the degradation of important cell proteins by the proteasome after cellular damage and, therefore, for contributing to cell protection. Heat shock proteins make a large family of vital proteins specialized in cell protection. They are present in all types of cells and are highly conserved across prokaryotes, plants and animals. They display a large range of molecular weights on which their classification is based: large (> 90 kDa), medium (30-90 kDa) and small (<30 kDa) HSP (review by Otaka et al., 2006). Prototypic proteins of these classes include HSP 90, HSP 70 and HSP 25 (or HSP 72 and HSP 27, depending on the animal species considered). They exist in the cell constitutively or are induced quickly in response to fast environmental changes. Following a stress, HSP expression is regulated through the heat shock factor HSF (HSF1 in particular), probably following cell accumulation of altered or aggregated cellular proteins (Otaka et al., 2006). Then HSF1 is phosphorylated and forms a trimer which translocates into the nucleus. This complex binds to the heat shock element (HSE) which in turn up-regulates HSP gene expression and protein synthesis. HSP and denatured/aggregated protein form complexes, thus preventing protein degradation, inhibiting HSF1 synthesis and trimerization, and down-regulating HSP gene activation.

Among the various HSP that have been characterized in the GIT, HSP 25 and HSP 70 are directly implicated in epithelial cell protection (Otaka et al., 2006; Petrof et al., 2004). HSP 25 is associated with the actin cytoskeleton and contributes to stabilize cell-cell contacts, including tight junctions. HSP 70 is more involved in intra-cellular protein chaperoning. Gastrointestinal HSP 25 and HSP 70 are normally produced in surface epithelial cells in contact with a low pH and a high proteolytic activity as in the stomach or with bacteria and fermentation products (e.g. short chain fatty acids like butyrate) in the colon and the distal ileum (Petrof et al., 2004). By contrast, duodenal and jejunal

Table 3. Main cytokines, their cell origin, effects on intestinal barrier function and tight junction proteins and signalling pathways involved (adapted from Al-Sadi et al., 2009).

Cytokine	Source	Barrier function[1]	Tight junction changes[2]	Intra-cellular signalling[3]
IFNγ	Lymphocytes, dendritic cells, monocytes	TEER (-) PCP (+)	Pinocytosis of TJ proteins (occludin, claudin-1) Claudin-2 (-)	Rho kinase, MLC phosphorylation
TNFα	Activated lymphocytes and macrophages	TEER (-) PCP (+)	TJ strand numbers and depths (-)	NFκB, MLCK, PKA
IL-1β	Various immuno-modulatory cells	TEER (-) PCP (+)	Occludin (-)	NFκB, MLCK
IL-4	CD4+ T lymphocytes	TEER (-) PCP (+)	Claudin-2 (-)	PI3K
IL-6	Various cells (incl. IEC cells)	TEER (-/+) PCP (+/-)	?	?
IL-10	T cells stimulated by IL-4	TEER (0) PCP (0)	Protective against barrier and TJ changes induced by IFNγ	
IL-13		TEER (-) PCP (+)	Claudin-2 (-)	Akt
TGF-β		TEER (+) PCP (-)	Claudin-1 (+)	ERK or PKC

[1]TEER: trans-epithelial electrical resistance; PCP: para-cellular permeability.
[2]TJ: tight junction.
[3]Akt, protein kinase B; ERK, extracellular signal-related kinase; MLC, myosin light chain; MLCK, MLC kinase; NFκB, nuclear factor kappaB; PKA, protein kinase A; PKC, protein kinase C; PI3K, phosphatidyl-inositol 3-kinase; Rho kinase, Ras homolog kinase.

(+) increased; (-) decreased; ?, unidentified.

epithelial cells from villi express very low levels of HSP, as do crypt cells of the small and large intestines in physiological conditions.

GIT cell and tissue expression of HSP is determined by various signals received by IEC cells from the lumen and from the underlying lamina propria (Figure 2).

In addition to heat shock and toxicants (e.g. heavy metals) that are known to activate HSP expression, these signals include nutrients (e.g. glutamine; Ehrenfried *et al.*, 1995; Phanvijhitsiri *et al.*, 2006; Singleton and Wischmeyer, 2006; Wischmeyer *et al.*, 1997), fermentation products (e.g. butyrate; Arvans *et al.*, 2005; Ren *et al.*, 2001), many bacterial products (e.g. *Escherichia coli* LPS, N-formyl-L-methionyl-L-leucyl-L-phenylalanine (fMLP) peptide, flagellin from *Salmonella*; Carlson *et al.*, 2007; Kojima *et al.*, 2004; Musch *et al.*, 2004; Petrof *et al.*, 2008) and inflammatory mediators (e.g. TNFα, IL-1β, IL-2, IL-10, IL-11) resulting from the activation of various cell types in the lamina propria (Kojima *et al.*, 2003; Ropeleski *et al.*, 2003).

Most of these factors have been demonstrated to confer gut protection by inducing HSP. HSP 25 is particularly involved in epithelial barrier strengthening, acting as an actin polymerization modulator and being

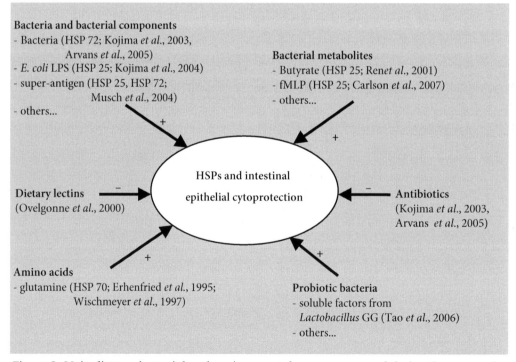

Figure 2. Main dietary, bacterial and environmental components modulating the expression of heat shock proteins (HSP) in IEC cells.

required for occludin up-regulation (Dalle-Donne *et al.*, 2001; Dokladny *et al.*, 2006; Panasenko *et al.*, 2003). HSP-induced cell protection often involves the down-regulation of nuclear factor kappa B (NF-kappaB), leading to inhibition of inflammatory cytokine production and induction of epithelial cell proliferation (Malago *et al.*, 2002). Finally, other factors, including antibiotics (Kojima *et al.*, 2003; 2004) and some plant lectins (Ovelgonne *et al.*, 2000) reduce both HSP expression and gut epithelial cell protection.

Gut homeostasis and the microbiota

Bacteria (both commensal and pathogenic) are determinant in the development of the gut and the maintenance of its homeostasis, particularly via the so-called toll-like receptors (TLR) present on gut epithelial cells (reviews by Abreu *et al.*, 2005, Cario *et al.*, 2005; Rakoff-Nahoum *et al.*, 2004) (Table 4).

Components of gut bacteria stimulate the expression of TJ proteins and participate in gut protection from injury, in addition to inducing digestive and metabolic enzymes. Various bacterial components, signalling pathways and effects on gut epithelial cells have been identified (Table 5).

Table 4. Known TLRs and their ligands (adapted from Cario et al., 2005).

TLRs	Ligands
TLR1, TLR2, TLR6	Lipoproteins
TLR3	Double strand RNA (dsRNA)
TLR4	Lipopolysaccharide (LPS)
TLR5	Flagellin
TLR7, TLR8	Single strand RNA (ssRNA)
TLR9	Cytosine-phosphate-guanine dinucleotides (CpG)
TLR11	Uropathogenic ligand

Table 5. Components of probiotic bacteria involved in gut epithelial modulation (adapted from Lebeer et al., 2008).

Bacterial strain[1]	Component[2]	Signalling[3]	Effect
Cell surface factors			
L. johnsonii	LTA La1	?	cytokine (IL-8) release
L. acidophilus	LTA La10	?	cytokine (IL-8) release
L. acidophilus ATCC 4356	?	?	anti-apoptotic, anti-inflammatory
L. rhamnosus GG	?	MAPK, Akt	anti-apoptotic
L. reuteri	?	IkB, EGF	
Secreted proteins			
L. rhamnosus GG	p40, p75	Akt, PI3K	epithelial homeostasis
L. casei ATCC 334	p40, p75	Akt, PI3K	cell growth, anti-apoptotic
L. casei ATCC 393	p40, p75	PKC, MAPK	cell growth, anti-apoptotic
Soluble peptides			
L. rhamnosus GG	?	MAPK, JNK	HSP induction, cell protection
L. rhamnosus GG	CSF	MAPK, Akt	HSP induction, cell protection
L. plantarum	CSF	MAPK, Akt	HSP induction, cell protection
Bacillus subtilis	CSF	MAPK, Akt	HSP induction, cell protection

[1] *L., Lactobacillus.*
[2] CSF, competence and sporulating factor; LTA, lipoteichoic acid; p40, p75, proteins of molecular weights 40 and 75 kDa; ?, unidentified.
[3] Akt, protein kinase B; EGF, epithelial growth factor; JNK, c-Jun N-terminal protein kinase; MAPK, mitogen-activated protein kinase; PI3K, phosphatidyl-inositol 3-kinase; PKC, protein kinase C; ?, unidentified.

3. Gut physiology in pigs around weaning: recent findings and implications for rearing practice

Recent findings

One hallmark of weaning is gut anatomical and functional alterations (Lallès *et al.*, 2004, 2007; Montagne *et al.*, 2007; Pluske *et al.*, 1997). These disorders include villous atrophy and enhance epithelial barrier permeability and chloride secretion (Boudry *et al.*, 2004; Spreeuwenberg *et al.*, 2001). They favour diarrhoea, translocation of bacteria and uptake of undesirable components from the diet (e.g. protein antigens) and from the microbiota (e.g. LPS and toxins). The transient anorexia and subsequent nutrient shortage often observed after weaning are probably the main inducers of these changes. However, recent data point to the importance of the weaning stress and the involvement of the nervous system in these disorders. Moeser *et al.* (2007a,b) demonstrated that both gut hyper-secretion and hyper-permeability resulted from the activation of the enteric nervous system and implicate the CRF pathway and mast cells. CRF receptor 1 (CRF-R1) is over-expressed in the jejunum and the colon of weaned pigs and the administration of a CRF receptor antagonist abolishes gut epithelial physiology disorders (Moeser *et al.*, 2007b). Although mucosal mast cell densities are reduced, mast cell degranulation and tryptase mucosal levels are greater in weaned than in non-weaned pigs (Moeser *et al.*, 2007a). Pharmacological stabilisation of mast cells abolishes the observed post-weaning gut physiology disorders. Besides, transient mucosal inflammation along the GIT after weaning has also been documented (McCracken *et al.*, 1999; Pié *et al.*, 2004). In particular, early responses were characterised by the up-regulation of inflammatory cytokine (IL-1β, IL-6 and TNF-α) genes. These changes, reflecting the activation of various cell types in mucosal tissues may also contribute to the observed gut permeability disorders. Collectively, the data obtained in pigs fit well those with rodent animals under stress. Finally, little is known on HSP in weaned pigs, apart from the observation that their gut longitudinal profiles vary largely following weaning (David *et al.*, 2002).

Implications for rearing practice

The recent discoveries on the importance of psychological stress on gut secretory physiology and permeability disorders strongly suggest that reducing the level of stress in weaner pigs may be a strategy to reduce post-weaning gut disorders. Besides lowering general management stress, delaying the age at weaning, stimulating feed intake and adapting the weaning diet composition are the main ways for achieving a low level of stress.

Pigs weaned at 19 days of age display more gut physiology alterations as described above than those weaned at 28 days (Moeser *et al.* 2007a,b). The level of feed intake is a major determinant of gut anatomical and functional disorders observed after weaning (reviews by Lallès *et al.*, 2004, 2007; Pluske *et al.*, 1997; van Beers-Schreurs *et al.*, 1998). Many nutritional strategies have been developed over the last decades for improving feed intake and minimizing gut alterations in weaned pigs. The most effective ones include supplementation of the diets with various organic acids used alone or as blends, specific amino acids and glutamine or spray-dried animal plasma protein (Lallès *et al.*, 2007, 2009). Reviewing these nutritional strategies is out of scope of this paper. However, two examples are provided and discussed in the light of gut barrier function and cell protection.

The first example is zinc (provided as oxide or in other chemical forms). It has been used over decades in young pigs at doses (1,000 to 3,000 mg/kg feed) well above requirements for reducing post-weaning diarrhoea and gut disorders. First, zinc requirements must be covered because zinc deprivation alters epithelial barrier integrity by disrupting tight and adherens junctions at the levels of ZO-1 and occludin, and by disorganising actin filaments (Finamore *et al.*, 2008). Recent findings strongly suggest that high levels of zinc trigger many mechanisms of gut protection that may account for its high efficacy. Zinc increases glucose absorption capacity and stimulates pancreatic and intestinal enzyme production (Carlson *et al.*, 2004; Hedemann *et al.*, 2006), thus favouring nutrient intestinal absorption and provision to the organism. Zinc inhibits secretagogue-induced chloride secretion of the small intestine via a direct effect on epithelial cells (Carlson *et al.*, 2004, 2006, 2008; Feng *et al.*, 2006). It also increases mRNA and protein expression of ZO-1 and occludin in pig ileal mucosa, thus improving epithelial barrier function at the TJ level (Zhang and Guo, 2009). Zinc reduces the intestinal expression of the stem cell factor (SCF, responsible for mast cell proliferation in the gut), mucosal mast cell numbers and histamine release (Ou *et al.*, 2007). In a context of infection, zinc is also known to inhibit the adhesion and the invasion of enterotoxigenic *E. coli* (ETEC) and to restore an optimal anti-inflammatory cytokine balance following ETEC infection of IEC lines (Roselli *et al.* 2003). One could anticipate that zinc supplementation may stimulate gut HSP expression and epithelial cell protection in pigs because zinc compounds (e.g. polaprezinc, zinc-L-carnosine) stimulate HSP production in IEC cells and in mice after oxidative or inflammatory stress (Ohkawara *et al.*, 2005, 2006). Although the detrimental impact of zinc on the environment cannot be ignored, its multiple protective effects on host gut physiology may serve as a basis for developing effective alternatives to in-feed antibiotics in pigs.

The second example deals with probiotic bacteria that have a potential for maintaining gut homeostasis (see previous sections). *E. coli* Nissle 1917 strain was able to abolish diarrhoea, through a reduction in jejunal secretagogue-induced chloride secretion and to restore alterations in paracellular permeability induced by ETEC challenge in weaned pigs (Schroeder *et al.*, 2006). The porcine *Lactobacillus sobrius* probiotic was recently shown to reduce ETEC adhesion to porcine IEC lines, to inhibit TJ ZO-1 delocalization, and to reduce occludin tissue concentration and dephosphorylation, and the rearrangement of actin filaments, thus preventing epithelial barrier damage induced by ETEC (Roselli *et al.*, 2007). It is remarkable that these effects were highly strain-specific. *L. sobrius* reduced *E. coli* infection and promoted the growth of infected pigs (Konstantinov *et al.*, 2008). Interstingly, a probiotic mixture was able to limit gut disorders in a swine model of necrotizing enterocolitis (Siggers *et al.*, 2008). Although still scarce, the results available in pigs illustrate that appropriate strains of probiotic bacteria can exert beneficial influences on gut physiology and barrier function after weaning. This promising area of research should expand in the future.

4. Influence of early life events on gut physiology

Antibiotic therapy and stress early in life are two factors able to alter the gut barrier function durably as shown in rodents. They seem to interfere with early bacterial colonisation of the gut. Maternal and offspring nutrition may contribute to maintain or restore gut barrier properties.

Neonatal stress

Early life stress, as repeated maternal separation of rodent pups can affect gut function immediately but also durably (Barreau *et al.*, 2004b; Gareau *et al.*, 2008; Soderholm and Perdue, 2006; Soderholm *et al.*, 2002). Short-term consequences of early stress include high colonic permeability to macromolecules and increased gut tissue attachment and translocation of commensal bacteria to extra-GIT tissues. Long-term issues are gut barrier disorders in response to mild stress, impaired host resistance to luminal bacteria and increased sensitivity to intestinal infestation by nematodes (Barreau *et al.*, 2006; Gareau *et al.*, 2006; Soderholm *et al.*, 2002). Alterations in nerve-mast cell interactions and in CRF and nerve growth factor (NGF)-mediated cholinergic regulation of gut barrier function were demonstrated (Barreau *et al.*, 2004a, 2007, 2008; Gareau *et al.*, 2007b). One important question raised by these studies is the cross-talk between the central nervous system and the gut microbiota (Collins and Bercik, 2009). To the best of our knowledge, no such information is available in pigs.

Early antibiotic therapy

Neonatal bacterial colonisation of the gut can be altered by broad-spectrum antibiotics. Such treatment has been shown to alter gastrointestinal developmental gene expression patterns, intestinal barrier transcriptome, and epithelial barrier function in adult rodents (Fak *et al.*, 2008b; Schumann *et al.*, 2005). In addition, gut mucosal mast cell densities and mast cell protease concentrations were increased (Nutten *et al.*, 2007). Although precise bacteria or bacterial components responsible for these alterations have yet to be identified, intestinal TLR like TLR-2 may be involved since it regulates barrier function and controls gut inflammation (Cario *et al.*, 2007).

Gut physiology alterations and nutrition

Feeding gestation diets enriched with n-3 poly-unsaturated fatty acids to sows result in an improvement of intestinal glucose uptake via an up-regulation of the glucose cotransporter SGLT-1, and increased energy storage in weaned pigs (Gabler *et al.*, 2007, 2009). Offspring sensitivity to gut permeability alterations induced by mast cell degranulation is also reduced (Boudry *et al.*, 2009). These improvements may be mediated, at least partly via changes in lipid composition of cell membranes.

Probiotic bacteria (lactobacilli) used alone or in combination with specific dietary ingredients (e.g. oligosaccharides, poly-unsaturated fatty acids) are able to alleviate gut barrier disorders observed in adult rodents submitted to neonatal stress (Eutamene *et al.*, 2007; Garcia-Rodenas *et al.*, 2006; Gareau *et al.*, 2007a). Responses to probiotic treatments are greater in young than in older pups, emphasizing the importance of the window of gut bacterial colonisation in this process (Fak *et al.*, 2008a).

5. Conclusions and perspectives

Major progress in the understanding of gut barrier function and epithelial cell protection and their regulation has been made over the last decade. Tight junctions and heat shock proteins (especially HSP 25 and HSP 70) appear to have a major role in gut defence. Experimental evidence from rodent models indicates that these defence systems are highly dynamic, integrated and finely regulated at the epithelial monolayer. The first level of gut barrier regulation is a direct cross-talk between the host and the gut microbiota. The second regulatory level involves the nervous system, mucosal mast cells and various mediators acting on the epithelium. Gut barrier function and its regulation can be altered following stress or after disturbing the microbiota (= dysbiosis). These findings have very recently led to the concept that the microbiota are an integrate component of the brain-gut axis: they provide the brain with luminal information while they are influenced by (or under control of?) the brain (Collins and Bercik, 2009). Adverse events (e.g. stress, antibiotic therapy) occurring early in life may have negative consequences on gut barrier function in adulthood, emphasizing the importance of gut bacterial colonisation during the neonatal period in the subsequent development of an optimal gut physiology. Interestingly, the first available data suggest that such long-term alterations in gut barrier function may be alleviated by specific strains of probiotic bacteria used alone or in combination with specific nutrients (e.g. poly-unsaturated fatty acids). Studies are still scarce in the swine species and, therefore, merit further developments in the areas of gut barrier function, epithelial cell protection and their regulations. This in turn, will help developing appropriate strategies for preventing gut disorders, including those usually observed after weaning. Long-term issues should also be considered.

Acknowledgements

Dr P. Bikker (Schothorst Feed Research, Lelystad, The Netherlands), Dr P. Leterme (Prairie Swine Center, Saskatoon, Canada) and Dr I.P. Oswald (INRA, Pharmacology & Toxicology, Toulouse, France) are greatly acknowledged for their critical reading of this review.

References

Al-Sadi, R., Boivin, M., Ma, T., 2009. Mechanism of cytokine modulation of epithelial tight junction barrier. Frontiers in Bioscience 14: 2765-2778.

Arvans, D.L., Vavricka, S.R., Ren, H., Musch, M.W., Kang, L., Rocha, F.G., Lucioni, A., Turner, J.R., Alverdy, J., Chang, E.B., 2005. Luminal bacterial flora determines physiological expression of intestinal epithelial cytoprotective heat shock proteins 25 and 72. American Journal of Physiology, Gastrointestinal and Liver Physiology 288: G696-G704.

Abreu, M.T., Fukata, M., Arditi, M., 2005. TLR signaling in the gut in health and disease. Journal of Immunology 174: 4453-4460.

Barreau, F., Cartier, C., Ferrier, L., Fioramonti, J., Bueno, L., 2004a. Nerve growth factor mediates alterations of colonic sensitivity and mucosal barrier induced by neonatal stress in rats. Gastroenterology 127: 524-534.

Barreau, F., Ferrier, L., Fioramonti, J., Bueno, L., 2004b. Neonatal maternal deprivation triggers long term alterations in colonic epithelial barrier and mucosal immunity in rats. Gut 53: 501-506.

Barreau, F., de Lahitte, J.D., Ferrier, L., Frexinos, J., Bueno, L., Fioramonti, J., 2006. Neonatal maternal deprivation promotes *Nippostrongylus brasiliensis* infection in adult rats. Brain Behavour & Immunology 20:254-260.

Barreau, F., Cartier, C., Leveque, M., Ferrier, L., Moriez, R., Laroute, V., Rosztoczy, A., Fioramonti, J., Bueno, L., 2007. Pathways involved in gut mucosal barrier dysfunction induced in adult rats by maternal deprivation: corticotrophin-releasing factor and nerve growth factor interplay. Journal of Physiology 580: 347-356.

Barreau, F., Salvador-Cartier, C., Houdeau, E., Bueno, L., Fioramonti, J., 2008. Long-term alterations of colonic nerve-mast cell interactions induced by neonatal maternal deprivation in rats. Gut 57: 582-590.

Boudry, G., Péron V., Le Huërou-Luron I., Lallès J.P., Sève, B. 2004. Weaning induces both transient and long-lasting modifications of absorptive, secretory, and barrier properties of piglet intestine. Journal of Nutrition 134: 2256-2262.

Boudry, G., Douard, V., Mourot, J., Lallès, J.P., Le Huërou-Luron, I., 2009. Linseed oil in the maternal diet during gestation and lactation modifies fatty acid composition, mucosal architecture, and mast cell regulation of the ileal barrier in piglets. Journal of Nutrition 139: 1110-1117.

Cario, E., 2005. Bacterial interactions with cells of the intestinal mucosa: TOLL-like receptors and NOD2. Gut 54: 1182-1193.

Cario, E., Gerken, G., Podolsky, D.K., 2007. Toll-like receptor 2 controls mucosal inflammation by regulating epithelial barrier function. Gastroenterology 132: 1359-1374.

Carlson, D., Poulsen, H.D., Sehested, J. 2004. Influence of weaning and effect of post weaning dietary zinc and copper on electrophysiological response to glucose, theophylline and 5-HT in piglet small intestinal mucosa. Comparative Biochemistry & Physiology A 137: 757-765.

Carlson, D., Sehested J., Poulsen, H.D., 2006. Zinc reduces the electrophysiological responses *in vitro* to basolateral receptor mediated secretagogues in piglet small intestinal epithelium. Comparative Biochemistry & Physiology A 144: 514-519.

Carlson, D., Sehested J., Feng, Z., Poulsen, H.D., 2008. Serosal zinc attenuate serotonin and vasoactive intestinal peptide induced secretion in piglet small intestinal epithelium *in vitro*. Comparative Biochemistry & Physiology A 149: 51-58.

Carlson, R.M., Vavricka, S.R., Eloranta, J.J., Musch, M.W., Arvans, D.L., Kles, K.A., Walsh-Reitz, M.M., Kullak-Ublick, G.A., Chang, E.B., 2007. fMLP induces Hsp27 expression, attenuates NF-kappaB activation, and confers intestinal epithelial cell protection. American Journal of Physiology, Gastrointestinal and Liver Physiology 292: G1070-G1078.

Collins, S.M., Bercik, P., 2009. The relationship between intestinal microbiota and the central nervous system in normal gastrointestinal function and disease. Gastroenterology 136: 2003-2014.

Dalle-Donne, I., Rossi, R., Milzani, A., Di Simplicio, P., Colombo, R., 2001. The actin cytoskeleton response to oxidants: from small heat shock protein phosphorylation to changes in the redox state of actin itself. Free Radical Biology & Medicine 31: 1624-1632.

David, J.C., Grongnet, J.F., Lallès, J.P. 2002. Weaning affects the expression of heat shock proteins in different regions of the gastrointestinal tract of piglets. Journal of Nutrition 132: 2551-2561.

Dokladny, K., Moseley, P.L., Ma T.Y., 2006. Physiologically relevant increase in temperature causes an increase in intestinal epithelial tight junction permeability. American Journal of Physiology, Gastrointestinal and Liver Physiology 290: G204-G212.

Ehrenfried, J.A., Chen, J., Li, J., Evers, B.M., 1995. Glutamine-mediated regulation of heat shock protein expression in intestinal cells. Surgery 118: 352-357.

Eutamene, H., Lamine, F., Chabo, C., Theodorou, V., Rochat, F., Bergonzelli, G.E., Corthésy-Theulaz, I., Fioramonti, J., Bueno, L., 2007. Synergy between *Lactobacillus paracasei* and its bacterial products to counteract stress-induced gut permeability and sensitivity increase in rats. Journal of Nutrition 137: 1901-1907.

Fak, F., Ahrne S., Linderoth A., Molin, G., Jeppsson, B., Weström, B., 2008a. Age-related effects of the probiotic bacterium *Lactobacillus plantarum* 299v on gastrointestinal function in suckling rats. Digestive Disease & Science 53: 664-671.

Fak, F., Ahrne S., Molin, G., Jeppsson, B., Weström, B., 2008b. Microbial manipulation of the rat dam changes bacterial colonization and alters properties of the gut in her offspring. American Journal of Physiology, Gastrointestinal and Liver Physiology 294: G148-G154.

Feng, Z., Carlson, D., Poulsen, H.D., 2006. Zinc attenuates forskolin-stimulated electrolyte secretion without involvement of the enteric nervous system in small intestinal epithelium from weaned piglets. Comparative Biochemistry & Physiology A 145: 328-333.

Finamore, A., Massimi, M., Conti Devirgiliis, L., Mengheri, E., 2008. Zinc deficiency induces membrane barrier damage and increases neutrophil transmigration in Caco-2 cells. Journal of Nutrition 138: 1664-1670.

Gabler, N.K., Spencer, J.D., Webel, D.M., Spurlock, M.E., 2007. *In utero* and postnatal exposure to long chain (n-3) PUFA enhances intestinal glucose absorption and energy stores in weanling pigs. Journal of Nutrition 137: 2351-2358.

Gabler, N.K., Radcliffe, J.S., Spencer, J.D., Webel, D.M., Spurlock, M.E., 2009. Feeding long-chain n-3 polyunsaturated fatty acids during gestation increases intestinal glucose absorption potentially via the acute activation of AMPK. Journal of Nutritional Biochemistry 20: 17-25.

Gallois, M., Rothkötter, H.J., Bailey, M., Stokes, C.R., Oswald, I.P., 2009. Natural alternatives to in-feed antibiotics in pig production: Can immunomodulators play a role? Animal 3:1644-1661.

García-Rodenas, C.L., Bergonzelli, G.E., Nutten, S., Schumann, A., Cherbut, C., Turini, M., Ornstein, K., Rochat, F., Corthésy-Theulaz, I., 2006. Nutritional approach to restore impaired intestinal barrier function and growth after neonatal stress in rats. Journal of Pediatrics, Gastroenterology & Nutrition 43: 16-24.

Gareau, M.G., Jury, J., Yang, P.C., MacQueen, G., Perdue, M.H., 2006. Neonatal maternal separation causes colonic dysfunction in rat pups including impaired host resistance. Pediatric Research 59: 83-88.

Gareau, M.G., Jury, J., MacQueen, G., Sherman, P.M., Perdue, M.H., 2007a. Probiotic treatment of rat pups normalises corticosterone release and ameliorates colonic dysfunction induced by maternal separation. Gut 56: 1522-1528.

Gareau, M.G., Jury, J., Perdue, M.H., 2007b. Neonatal maternal separation of rat pups results in abnormal cholinergic regulation of epithelial permeability. American Journal of Physiology, Gastrointestinal and Liver Physiology 293: G198-G203.

Gareau, M.G., Silva, M.A., Perdue, M.H., 2008. Pathophysiological mechanisms of stress-induced intestinal damage. Current Molecular Medicine 8: 274-281.

Harhaj, N.S., Antonetti, D.A., 2004. Regulation of tight junctions and loss of barrier function in pathophysiology. International Journal of Biochemistry & Cell Biology 36: 1206-1237.

Hedemann, M.S., Jensen, B.B., Poulsen, H.D., 2006. Influence of dietary zinc and copper on digestive enzyme activity and intestinal morphology in weaned pigs. Journal of Animal Science 84: 3310-3320.

Kojima, K., Musch, M.W., Ren, H., Boone, D.L., Hendrickson, B.A., Ma, A., Chang, E.B., 2003. Enteric flora and lymphocyte-derived cytokines determine expression of heat shock proteins in mouse colonic epithelial cells. Gastroenterology 124: 1395-1407.

Kojima, K., Musch, M.W., Ropeleski, M.J., Boone, D.L., Ma, A., Chang, E.B., 2004. *Escherichia coli* LPS induces heat shock protein 25 in intestinal epithelial cells through MAP kinase activation. American Journal of Physiology, Gastrointestinal and Liver Physiology 286: G645-G652.

Konstantinov, S.R., Smidt, H., Akkermans, A.D., Casini, L., Trevisi, P., Mazzoni, M., De Filippi, S., Bosi, P., de Vos, W.M., 2008. Feeding of *Lactobacillus sobrius* reduces *Escherichia coli* F4 levels in the gut and promotes growth of infected piglets. FEMS Microbiology & Ecology; 66: 599-607.

Lallès, J.P., Boudry, G., Favier, C., Le Floc'h, N., Luron, I., Montagne, L., Oswald, I.P., Pié, S., Piel, C., Sève, B., 2004. Gut function and dysfunction in young pigs: physiology. Animal Research 53: 301-316.

Lallès, J.P., Bosi P., Smidt H., Stokes, C.R., 2007. Nutritional management of gut health in pigs around weaning. Proceedings of the Nutrition Society 66: 260-268.

Lallès J.P., Bosi P., Janczyk P., Koopmans S.J., Torrallardona D., 2009. Impact of bioactive substances on the gastrointestinal tract and performance of weaned piglets: a review. Animal 3: 1625-1643.

Lebeer, S., Vanderleyden, J., De Keersmaecker, S.C., 2008. Genes and molecules of lactobacilli supporting probiotic action. Microbiology & Molecular Biology Reviews 72: 728-764.

Ma, T.Y., Anderson, J.M., 2006. Tight junctions and intestinal barrier. In: Johnson, L.R., (ed.) Physiology of the gastrointestinal tract. Academic Press, New York, USA, Volume 1, pp. 1559-1594.

Malago, J.J., Koninkx, J.F., Van Dijk, J.E., 2002. The heat shock response and cytoprotection of the intestinal epithelium. Cell Stress Chaperones 7: 191-199.

McCracken, B.A., Spurlock, M.E., Roos, M.A., Zuckermann, F.A., Gaskins, H.R., 1999. Weaning anorexia may contribute to local inflammation in the piglet small intestine. Journal of Nutrition 129: 613-619.

McKay, D.M., Baird, A.W., 1999. Cytokine regulation of epithelial permeability and ion transport. Gut 44: 283-289.

Moeser, A.J., Klok, C.V., Ryan, K.A., Wooten, J.G., Little, D., Cook, V.L., Blikslager, A.T., 2007a. Stress signaling pathways activated by weaning mediate intestinal dysfunction in the pig. American Journal of Physiology, Gastrointestinal and Liver Physiology 292: G173-G181.

Moeser, A.J., Ryan, K.A., Nighot, P.K., Blikslager, A.T., 2007b. Gastrointestinal dysfunction induced by early weaning is attenuated by delayed weaning and mast cell blockade in pigs. American Journal of Physiology, Gastrointestinal and Liver Physiology 293: G413-G421.

Montagne, L., Boudry, G., Favier, C., Le Huërou-Luron, I., Lallès, J.P., Sève, B., 2007. Main intestinal markers associated with the changes in gut architecture and function in piglets after weaning. British Journal of Nutrition 97: 45-57.

Musch, M.W., Petrof, E.O., Kojima, K., Ren, H., McKay, D.M., Chang, E.B., 2004. Bacterial superantigen-treated intestinal epithelial cells upregulate heat shock proteins 25 and 72 and are resistant to oxidant cytotoxicity. Infection & Immunity 72: 3187-3194.

Nutten, S., Schumann, A., Donnicola, D., Mercenier, A., Rami, S., Garcia-Rodenas, C.L., 2007. Antibiotic administration early in life impairs specific humoral responses to an oral antigen and increases intestinal mast cell numbers and mediator concentrations. Clinical & Vaccine Immunology 14: 190-197.

Ohkawara, T., Takeda, H., Kato, K., Miyashita, K., Kato, M., Iwanaga, T., Asaka, M., 2005. Polaprezinc (N-(3-aminopropionyl)-L-histidinato zinc) ameliorates dextran sulfate sodium-induced colitis in mice. Scandinavian Journal of Gastroenterology 40: 1321-1327.

Ohkawara, T., Nishihira, J., Nagashima, R., Takeda, H., Asaka, M., 2006. Polaprezinc protects human colon cells from oxidative injury induced by hydrogen peroxide: relevant to cytoprotective heat shock proteins. World Journal of Gastroenterology 12: 6178-6181.

Ou, D., Li, D., Cao, Y., Li, X., Yin, J., Qiao, S., Wu, G., 2007. Dietary supplementation with zinc oxide decreases expression of the stem cell factor in the small intestine of weanling pigs. Journal of Nutritional Biochemistry 18: 820-826.

Ovelgonne, J.H., Koninkx, J.F., Pusztai, A., Bardocz, S., Kok, W., Ewen, S.W., Hendriks, H.G., Van Dijk, J.E., 2000. Decreased levels of heat shock proteins in gut epithelial cells after exposure to plant lectins. Gut 46: 679-687.

Otaka, M., Odashima, M., Watanabe, S., 2006. Role of heat shock proteins (molecular chaperones) in intestinal mucosal protection. Biochemical and Biophysical Research Communications 348: 1-5.

Palazzo, M., Gariboldi, S., Zanobbio, L., Selleri, S., Dusio, G.F., Mauro, V., Rossini, A., Balsari, A., Rumio, C., 2008. Sodium-dependent glucose transporter-1 as a novel immunological player in the intestinal mucosa. Journal of Immunology 181: 3126-3136. Erratum in: Journal of Immunology 181: 7428.

Panasenko, O.O., Kim, M.V., Marston, S.B., Gusev, N.B., 2003. Interaction of the small heat shock protein with molecular mass 25 kDa (hsp25) with actin. European Journal of Biochemistry 270: 892-901.

Pappenheimer, JR., 1988. Physiological regulation of epithelial junctions in intestinal epithelia. Acta Physiologica Scandinavica Supplement 571:43-51.

Petrof, E.O., Ciancio, M.J., Chang, E.B., 2004. Role and regulation of intestinal epithelial heat shock proteins in health and disease. Chinese Journal of Digestive Disease 5: 45-50.

Petrof, E.O., Musch, M.W., Ciancio, M., Sun, J., Hobert, M.E., Claud, E.C., Gewirtz, A., Chang, E.B., 2008. Flagellin is required for salmonella-induced expression of heat shock protein Hsp25 in intestinal epithelium. American Journal of Physiology, Gastrointestinal and Liver Physiology 294: G808-G818.

Phanvijhitsiri, K., Musch, M.W., Ropeleski, M.J., Chang, E.B., 2006. Heat induction of heat shock protein 25 requires cellular glutamine in intestinal epithelial cells. American Journal of Physiology, Cell Physiology 291: C290-C299.

Pié, S., Lallès, J.P., Blazy, F., Laffitte, J., Sève, B., Oswald, I.P., 2004. Weaning is associated with an upregulation of expression of inflammatory cytokines in the intestine of piglets. Journal of Nutrition 134: 641-647.

Pluske, J.R., Hampson, D.J., Williams, I.H., 1997. Factors influencing the structure and function of the small intestine in the weaned pig: A review. Livestock Production Science 51: 215-236.

Rakoff-Nahoum, S., Paglino, J., Eslami-Varzaneh, F., Edberg, S., Medzhitov, R., 2004. Recognition of commensal microflora by toll-like receptors is required for intestinal homeostasis. Cell 118: 229-241.

Ren, H., Musch, M.W., Kojima, K., Boone, D., Ma, A., Chang, E.B., 2001. Short-chain fatty acids induce intestinal epithelial heat shock protein 25 expression in rats and IEC 18 cells. Gastroenterology 121: 631-639.

Ropeleski, M.J., Tang, J., Walsh-Reitz, M.M., Musch, M.W., Chang, E.B., 2003. Interleukin-11-induced heat shock protein 25 confers intestinal epithelial-specific cytoprotection from oxidant stress. Gastroenterology 124: 1358-1368.

Roselli, M., Finamore, A., Garaguso, I., Britti M.S.,, Mengheri, E., 2003. Zinc oxide protects cultured enterocytes from the damage induced by *Escherichia coli*. Journal of Nutrition 133: 4077-4082.

Roselli, M., Finamore, A., Britti, M.S., Konstantinov, S.R., Smidt, H., de Vos, W.M., Mengheri, E., 2007. The novel porcine *Lactobacillus sobrius* strain protects intestinal cells from enterotoxigenic *Escherichia coli* K88 infection and prevents membrane barrier damage. Journal of Nutrition 137: 2709-2716.

Schroeder, B., Duncker, S., Barth, S., Bauerfeind, R., Gruber, A.D., Deppenmeier, S., Breves, G., 2006. Preventive effects of the probiotic *Escherichia coli* strain Nissle 1917 on acute secretory diarrhea in a pig model of intestinal infection. Digestive Disease & Science 51: 724-731.

Schumann, A., Nutten, S., Donnicola, D., Comelli, E.M., Mansourian, R., Cherbut, C., Corthesy-Theulaz, I., Garcia-Rodenas, C., 2005. Neonatal antibiotic treatment alters gastrointestinal tract developmental gene expression and intestinal barrier transcriptome. Physiological Genomics, 23: 235-245.

Siggers, R.H., Siggers, J., Boye, M., Thymann, T., Mølbak, L., Leser, T., Jensen, B.B., Sangild, P.T., 2008. Early administration of probiotics alters bacterial colonization and limits diet-induced gut dysfunction and severity of necrotizing enterocolitis in preterm pigs. Journal of Nutrition 138: 1437-1444.

Singleton, K.D., Wischmeyer, P.E., 2006. Oral glutamine enhances heat shock protein expression and improves survival following hyperthermia. Shock 25: 295-299.

Soderholm, J.D., Yates, D.A., Gareau, M.G., Yang, P.C., MacQueen, G., Perdue, M.H., 2002. Neonatal maternal separation predisposes adult rats to colonic barrier dysfunction in response to mild stress. American Journal of Physiology, Gastrointestinal and Liver Physiology 283: G1257-G1263.

Soderholm, J.D., Perdue M.H., 2006. Effect of stress on intestinal mucosal function. In: Johnson, L.R. (Ed.), Physiology of the gastrointestinal tract. Academic Press, New York, USA, Volume 1, pp. 763-780.

Spreeuwenberg, M.A., Verdonk, J.M., Gaskins, H.R., Verstegen, M.W., Wang, J., Chen, L., Li, P., Li, X., Zhou, H., 2001. Small intestine epithelial barrier function is compromised in pigs with low feed intake at weaning. Journal of Nutrition 131: 1520-1527.

Tao, Y., Drabik, K.A., Waypa, T.S., Musch, M.W., Alverdy, J.C., Schneewind, O., Chang, E.B., Petrof, E.O., 2006. Soluble factors from *Lactobacillus* GG activate MAPKs and induce cytoprotective heat shock proteins in intestinal epithelial cells. American Journal of Physiology, Gastrointestinal and Liver Physiology 290: C1018-C1830. Erratum in: American Journal of Physiology, Gastrointestinal and Liver Physiology 291: C194.

Van Beers-Schreurs, H.M., Nabuurs, M.J., Vellenga, L., Kalsbeek-Van der Valk, H.J., Wensing, T., Breukink, H.J., 1998. Weaning and the weanling diet influence the villous height and crypt depth in the small intestine of pigs and alter the concentrations of short-chain fatty acids in the large intestine and blood. Journal of Nutrition 128: 947-953.

Wischmeyer, P.E., Musch, M.W., Madonna, M.B., Thisted, R., Chang, E.B., 1997. Glutamine protects intestinal epithelial cells: role of inducible HSP70. American Journal of Physiology 272: G879-G884.

Yu, L.C., Flynn, A.N., Turner, J.R., Buret, A.G., 2005. SGLT-1-mediated glucose uptake protects intestinal epithelial cells against LPS-induced apoptosis and barrier defects: a novel cellular rescue mechanism? FASEB Journal 19: 1822-1835.

Yu, L.C., Huang, C.Y., Kuo, W.T., Sayer, H., Turner, J.R., Buret, A.G., 2008. SGLT-1-mediated glucose uptake protects human intestinal epithelial cells against *Giardia duodenalis*-induced apoptosis. International Journal of Parasitology 38: 923-934.

Zhang, B., Guo, Y., 2009. Supplemental zinc reduced intestinal permeability by enhancing occludin and zonula occludens protein-1 (ZO-1) expression in weaning piglets. British Journal of Nutrition 102: 687-693.

Zyrek, A.A., Cichon, C., Helms, S., Enders, C., Sonnenborn, U., Schmidt, M.A., 2006. Molecular mechanisms underlying the probiotic effects of *Escherichia coli* Nissle 1917 involve ZO-2 and PKCzeta redistribution resulting in tight junction and epithelial barrier repair. Cell Microbiology 9: 804-816.

Effects of oxidative stress and selenium supplementation on piglets and the underlying mechanisms

Chen Daiwen, Yu Bing, Yuan Shibin, Zhang Keying and He Jun
Institute of Animal Nutrition, Key Laborotary of Animal Disease-Resistance Nutrition, Sichuan Agricultural University, Yaan, Sichuan 625014, P.R. China; Chendwz@sicau.edu.cn

Abstract

Reactive oxygen species (ROS), such as the superoxide anion (O_2^-), hydrogen peroxide (H_2O_2) and hydroxyl radical (HO^-), are highly unstable species with unpaired electrons, capable of initiating oxidative damage. Normally, the generation and scavenging of ROS in body keeps a physiological balance. However, some abnormalities such as disease, damage, transportation and air pollution as well as the chemical toxins will result in over accumulation of ROS, which surpasses the scavenging capability of the body. The disrupted balance is usually called oxidative stress which is closely associated with a series of dysfunctions (Li *et al.*, 2006).

The accumulation of ROS is believed to induce the peroxidative reactions of tissue components such as proteins, DNA and lipids, which subsequently lead to the oxidative damage. In addition, the detrimental effects of oxidative stress on food intake, growth and muscle gain has brought considerable economic loss for the livestock producers. Therefore, a valid approach of preventing animals from oxidative damage has attracted considerable research interests in the world (Wu and Hu, 2005). However, most of current studies about the underlying mechanisms of oxidative stress mainly focused on experimental animals such as the mice and rat, and little are known about the poultry and domestic animals. In this article, we will summarize our current knowledge of oxidative stress in weaned piglets. A better knowledge of the mechanisms will help in proposing new strategies to prevent animals from the oxidative damages.

1. Development of the animal models

At present most of researchers focus their attention on disease-induced oxidative damage. The experimental animals like the rats are usually used as the model for study. However, few or no models are set up on the basis of poultry and domestic animals. Thus, we summarized many existing models and successfully developed a research model for weaned piglets.

Induction of oxidative stress by oxidated fish oil

Plant and animal oil (i.e. fish oil) are rich in PUFA. Improper storage of these lipids usually induces the peroxidative reaction of lipids and forms a large number of peroxides. These products not only depress the digestibility of lipid but induce oxidative stress for animals. To illustrate the underlying mechanisms and prevent tissue damage, a replicable animal model from food origin is necessary to develop.

In our studies, we firstly compared the oxidation speed of different oils. Our results showed that the oxidation speed of fish oil was much faster than corn oil when exogenous oxidation accelerators (Fe^{2+}, Cu^{2+}, H_2O_2 and O_2) were added. Through correlation analysis, we obtained two dynamic equations that are suitable for the evaluation of peroxide value (POV) for corn oil ($Y_{POV} = 0.0005X^2 + 0.1434X + 4.9404$, $R^2 = 0.9924$, X: day, Y: $meqO_2/kg$) and fish oil ($Y_{POV} = -0.692X^2 + 73.605X - 329.88$, $R^2 = 0.79$, X: day, Y: $meqO_2/kg$), respectively.

To develop the animal model, the oxidated fish oils with different POV values were prepared. These oils were supplemented into piglet's diet at a level of 3%. Results indicated that these oil supplementations significantly increased the incidence of diarrhea, and decreased the performance and nutrient digestibility in weaned piglets. It's noteworthy that the fish oil with a POV value of 1000 $meqO_2/kg$ was much more effective to induce these detrimental effects. As to why these detrimental effects occurred after oil supplementation, one important reason is that the produced hyperoxides by lipid oxidation impaired the digestive organs and increased tissue and cell penetrability (Huang *et al.*, 1988), which ultimately increased the incidence of diarrhea and declined nutrient digestibility. The other reason is probably due to the influences of lipid hyperoxides on hypothalamus-pituitary-adrenal axis (HPAA). The elevated levels of blood cortin and catecholamine have resulted in metabolic breakdown of nutrients (hepatin, proteins and lipids), which ultimately depressed the growth performance. However, we failed to find a linear correlation between the extent of oxidative damage and POV . This is probably due to the altered peroxide content in fish oil. We found that the POV of fish oils was increased by 34, 60 and 90 fold after 25, 60 and 80 h oxidative reaction, respectively. However, the dietary POV was only increased by 1.49, 1.93 and 2.27 fold after supplementing the oils into the diet (compared with the fresh fish oil). Another reason could be attributed to the forming of a large number of hyperoxides during lipid oxidation. Previous study has indicated that the oxidation of PUFA in fish oil produced many short-chain fatty acids (SCFA), which probably alleviated the oxidative damages by enhancing the efficiency of nutrient utilization (Eder *et al.*, 2003). Thus, it is difficult to well and truly evaluate the effects of oxidative stress through an animal model that is developed by dietary supplementation of oxidated lipid. The crucial factor is that the great changes occurred after the lipid was supplemented into the diet and entered into the digestive tract. Therefore, a more maneuverable and replicable animal model is need to be developed for further study.

Comparative induction of oxidative stress by diquat and oxidated fish oil

Diquat (DQ) is a bipyridyl herbicide converting molecular oxygen into superoxide anion radical which subsequently forms hydrogen peroxide by redox cycling metabolism (Burk *et al.*, 1995). Since the main target organ of diquat is the liver in which most nutrient metabolisms occurred, the development of a model using diquat may be more suitable and effective to reveal the underlying mechanism of oxidative stress (Fu *et al.*, 1999). Previous study indicated that the LD_{50} of diquat for rats is about 120 mg/kg, and a rat model has been successfully developed by celiac injection of diquat at a dosage of 12 mg/kg (Farrington *et al.*, 1973). In our study, this dosage has been applied to piglets and vomit occurred within 30 mins after injection. The feed intake resumed after 3 days and none of these piglets died throughout the trial (26 d). However, most of piglets lost their weight appetite, and spirits in the late experimental period.

In our study, we successfully developed a piglet model by supplementing 5% of oxidated fish oil into the diet (POV before/after supplementation: 786.5/122.63 meqO$_2$/kg). Our results showed that both diquat and oxiated fish oil successfully induced oxidative stress in weaned piglets. Compared with the fish oil group, the activities of SOD and GPx in the liver of diquat group were decreased by 16.9% and 26.4%, respectively. However, the content of MDA was increased by 30.8%. Therefore, the diquat was more powerful than fish oil as the stress inducer. This is probably due to the activation of lipid peroxidizing by O_2^- and H_2O_2 (Burk *et al.*, 1995) which came from O_2 oxidation after diquat injection. The continuously produced lipid peroxides forms a series of closely linked chains and continues to accumulate, which ultimately depresses the body anti-oxidative capacity by consuming a considerable amount of anti-oxidative enzymes (Schonbaum and Chance, 1976; Skufca *et al.*, 2003). As a result, the physiological balance has been disrupted and oxidative damages (i.e. liver necrosis) occurred. Thus, our results suggested that the diquat can be used as a valid inducer for oxidative stress. However, the injection dosage of 12 mg/kg was not suitable for the development of a chronic animal model. Our latest result has indicated that the appropriate injection dosage for piglets is 10 mg/kg.

2. The detrimental effects of oxidative stress on weaned piglets

Previous studies indicated that the excessive accumulating of ROS has resulted in oxidative damages of lipid, protein and DNA, which subsequently depressed the immune ability and increased the susceptibility to various diseases (Wang and Jin, 2002). After a series of studies, we found that the detrimental effects of oxidative stress on weaned piglets are mainly focused on following aspects.

Decrease of performance and nutrient digestibility

Our previous study showed that oxidative stress has negative effects on ADG, ADFI and F/G in weaned piglets. Compared with the control piglets, the ADG and ADFI of stressed piglets injected with diquat at a dosage of 12 mg/kg were decreased by 31.5% and 23.7%, respectively. The feed efficiency enhanced by 11.6%. In addition, the dosage of 10 mg/kg has also been tried in this study. Compared with the control piglets, the ADG and ADFI of stressed piglets were decreased by 12.2% and 2.5%, respectively. However, the F/G was increased by 10.5% (Yuan *et al.*, 2007). Our results also clearly showed that the oxidative stress has negative effects on nutrients digestibility. Compared with the control piglets, the digestibilities of crude protein (CP), gross energy (GE),dry matter (DM) and ether extract(EE) for stressed piglets injected with 12 mg/kg diquat were decreased by 14.4%, 13.9%, 13.8% and 16.9%, respectively. However, a smaller injection dosage (10 mg/kg) has resulted in a lesser decline of digestibility for CP (9.8%), DM (4.0%) and GE (3.7%) (Yuan *et al.*, 2007).

Decreases of antioxidant capability in tissues

Normally, the generation and scavenging of ROS keeps a physiological balance (Tong, 2003). There are two systems being capable of scavenging ROS in tissues: one is the antioxidant enzyme system, including enzymes such as SOD, CAT and GPx, and the other is non-enzyme system, including Vitamin C, Vitamin E, GSH, carotenoids and trace minerals such as copper, zinc and selenium. The two systems are both effective to prevent animals from oxidative damages.

Compared with the control piglets, the liver and plasma SOD activities of diquat- injected piglets (12 mg/kg) were decreased by 40.7% and 29.2%, respectively. However, the activities of GSH-PX decreased by 41.3% and 37.8%, respectively. Compared with the piglets received a diet containing 0.2 mg/kg Se, the serum activities of GSH and CAT from diquat-injected piglets (10 mg/kg, feed with the same diet) were decreased by 20.7% and 18.1%, respectively. In addition, both the activities of SOD and GSH-PX in serum and intestinal mucosa were significantly decreased ($P<0.01$). However, both the activities of MDA and XOD acutely enhanced ($P<0.01$) (Yuan et al., 2007).

Our results indicated that the physiological balance of ROS system has been disturbed by oxidative stress. The reduced activity of the main antioxidant enzymes is probably due to the scavenging of ROS which consumed a large number of enzymes.

Changes of blood parameters

Oxygen-carrying cells erythrocytes are continuously exposed to oxidative stress in the circulation, and undergo various oxidative damages. The blood platelets are flattened biological cells resembling stiff oblate spheroids for stopping hemorrhage and repairing injured tissue by forming a blood clot. Leukocytes are nucleated cells in peripheral blood, and their quantity is significantly lower than erythrocytes. According to their morphology, the leukocytes can be classified into lymphocytes, neutrophils and monocytes. It's well known that leukocyte arrest, recruitment and subsequent rolling, activation, adhesion and transmigration are essential stages in the immune response system to inflammation.

Previous studies have showed that the oxidative stress significantly decreased the number of erythrocytes, leukocytes and lymphocytes. However, the number of neutrophils acutely increased (Tong, 2003) .Usually the anemia is featured by the reduction of red blood cell (RBC) and hemoglobin(HGB). This is probably due to the mobilizing of the antioxidant enzyme CAT which consumed a large number of irons. Another possible reason is that the oxidative stress decreases the nutrient digestibility in piglets, which causes the shortage of materials for RBC and HGB synthesis. In addition, the stressed cells produce a large number of ROS which damages the membrane and accelerates the aging of the erythrocytes (Cheng et al., 2005).

At present, we believed that the volume and quantity of the blood platelets are determined by megakaryocytes. There are many parameters such as the mean platelets volume (MPV), platelet distribution width (PDW) and platelet count (PLT) being widely used to assess the quality of the platelets (Zhao et al., 2001). Previous studies indicated that the reduction of blood platelets has followed by the declines of MPV and PDW. This is probably due to the pathological damage of microvascular (caused by oxidative stress), which consumed a large number of blood platelets.

The nuetrophils accounts for about 50% of the total white blood cells. Thus, the amounts of leukocytes are significantly affected by the quantity of neutrophils. Our results showed that the oxidative stress acutely increased the amounts of total leukocytes and neutrophils. The increased neutrophils is essential in the activating of non-specific immunity and plays a crucial role in defense and resistance to pathogen invasion (Xiong and Li, 2003).

Influences on cell and tissue morphology

Damage of membrane lipids

It is well known that PUFA is more sensitive to ROS than other substances. The oxidative damages are usually characterized by the forming of lipid hyperoxides. However, the most widely used method to evaluate the extent to which the lipid hyperoxides formed is to determine the content of MDA and TBARS. Our results showed that the tissue content of MDA significantly increased during the oxidative stress. Compared with the control piglets, the liver and plasma MDA contents of diqaut-injected piglets (12 mg/kg) were increased by 87.2% and 69.2%, respectively. Compared with the piglets received a diet containing 0.2 mg/kg Se, the serum MDA contents of diquat-injected piglets (10 mg/kg, feed with the same diet) were increased by 42.3% and 26.0%, respectively.

Cell apoptosis in weaned piglets

Apoptosis is the process of programmed cell death (PCD) that may occur in multicellular organisms. Programmed cell death involves a series of biochemical events leading to a characteristic cell morphology and death, in more specific terms, a series of biochemical events that lead to a variety of morphological changes, including blebbing, changes to the cell membrane such as loss of membrane asymmetry and attachment, cell shrinkage, nuclear fragmentation, chromatin condensation, and chromosomal DNA fragmentation. Usually, the quantity of the DNA fragments can be used as an index for the ratio of apoptosis. Our results showed that pathological changes of ultrastructures in liver and spleen occurred in stressed piglets. The main changes included nuclear membrane expansion, shrinkage, nuclear fragmentation, chromatin condensation, and mitochondrial swelling. The lesions of splenic lymphocytes and plasma and reticular cells also occurred in stressed piglets. Compared with the control piglets, the DNA fragment produced in splenic cells was more evident in stressed piglets and the agarose gel electrophoresis showed a marked DNA ladder. Our results suggested that diquat injection was effective to induce the oxidative stress and subsequently resulted in the damage of DNA (Kerr and Harmon, 1991). In this study, we also found that the DNA degradation in liver was higher than the spleen. This is probably due to the fact that liver is not only the principal organ for nutrient metabolisms, but also the target organ for diquat (Burk *et al.*, 1995).

Damage of the immune system

The cell-mediated immunity is an immune response that does not involve antibodies but rather involves the activation of macrophages, natural killer cells (NK), and antigen-specific cytotoxic T-lymphoctyes in response to antigens. However, the humoral immune response is mediated by antibodies produced in the cells of the B lymphocytes. Secreted antibodies bind to antigens on the surfaces of invading microbes (such as viruses or bacteria), which flags them for destruction. The humoral immune response is an important aspect of immunity that is closely related to the immunoglobulin concentration and antibody titers. Therefore, the influences of oxidative stress on aniaml humoral immunity can be evaluated by measuring the two parameters.

In our studies, we found that the serum immunoglobulin level and other immue parameters such as Ea, Et and BLP (bacterial lipoprotein) in stressed piglets significatnly decreased. In addition, the antidody titers increased more slowly in stressed piglets after the injection of hog cholera vaccine (Li *et al.*, 2007). Previous studies indicated that the oxidative stress is capable of inducing an immune suppression, not only delaying the lymphocyte proliferation, differentiation and maturation, but also suppressing the ability of NK and anti-(K)-cells againist infections (Galan *et al.*, 1997). Another study has showed that the oxidative stress directly induced the tissue atrophy in thymus and spleen, and suppressed the proliferation of the splenic T lymphocytes. In our study, we found that the reduction of lymphocytes was probably due to the oxidative damage in spleen (Li *et al.*, 2007). Our hypothesis can be easily proved by the occurrence of DNA ladder in splenocytes and the pathological changes in splenic ultrastructures.

Decrease of gene expression of digestive enzymes and glucose transporters in intestinal mucosa

Previous studies indicated that the oxidative stress damaged the morphology of intestinal mucosa and decreased the activity of many brush border enzymes (Ghishan *et al.*, 1990; Marchionatti *et al.*, 2001; Prabhu *et al.*, 2000). Like with the previous study, we found that both the BBMV and BLMV sucrase activities were significantly decreased in stressed piglets ($P<0.05$),while the activities of alkaline phosphatase (ALP) in BBMV and Na^+-K^+-ATPase in BLMV were acutely decreased in stressed piglets ($P<0.01$) (Li, personal communication). Thus, the decreased nutrient digestibility caused by oxidative stress probably results from the decreased activity of many digestive enzymes in intestinal mucosa. Zhang reported that the ischemia-induced oxidative stress significantly reduced the Na^+ permeability in intestinal brush-border. In our study, we also found the reduction of intestinal absorption of glucose and Na^+ in piglets subjected to oxidative stress. However, the absorption of glucose increased when the Na^+ concentration was enhanced. Compared with the control piglets, the transportations for glucose in BBMV and BLMV were both acutely decreased ($P<0.01$) in stressed piglets. Further study has indicated that the decreased glucose transportation (caused by 10 mg/kg injection of diquat) may be attributed to the down-regulation of glucose transporter SGLT1 and GLUT2 mRNA expressions (Li, personal communication). The Na^+/glucose co-transporter is an improtant membrane protein localized in the brush border of the intestinal epithelium that is responsible for the uptake of the dietary D-glucose and D-galactose from the intestinal lumen (Bird *et al.*, 1996; Tarpey *et al.*, 1995). The abundance of SGLT1 mRNA is closely associated with the villus height. Usually, the longer villi suggests the higher expression level of SGLT1 mRNA.

The glucose transporter 2 (GLUT2) is the major glucose transporter isoform expressed in hepatocytes, insulin-secreting pancreatic beta cells, and absorptive epithelial cells of the intestinal mucosa and kidney. It functions as a low affinity, but high-turnover transport system; together with the SGLT, it is thought to act as a glucose-sensing apparatus that plays a role in blood glucose homeostasis by responding to changes in blood glucose concentration and altering the rate of glucose uptake into liver cells (Guan and Liu, 2000). Many factors such as the dietary nutrients, hormones and growth factors are proved to be able to regulate the expression of SGLT1 and GLUT2 mRNA. Our latest results showed that oxidative stress acutely decreased the expression of intestinal SGLT1 and GLUT2 mRNA in weaned piglets ($P<0.01$). Previous study proved that the reduction of glucose tranporter expression (caused by oxidative stress) may activate the NF-kB channel through PKC pathway (Han *et al.*, 2000). However, David *et al* (2001) reported that the down-regulation of SGLT1 expression in BBMV may be attributed to the disorder of membrane fluidity caused by oxidative damage (Heuil and Meddings, 2001). As to the details how the oxidative stress decreased the abundance of SGLT1 and GLUT2 mRNA is still unclear. The underlying mechanisms should be illustrated by further researches.

3. Effects of selenium supplementation on oxidative stress

Usually, the produced ROS in animal body has been neutralized by anti-oxidative system. Besides meeting the normal requirements for animals, the nutrients in feed stuffs should be also enough to maintain the balance of ROS system (Fang *et al.*, 2003, 2004). That is to say that nutrient requirements for animals may be altered under oxidative stress. Selenium (Se) has previously been proved to be one of the essential microelements for animals. Burk and Hill (1995) has indicated that over 100 different selenoproteins existed in mammalian tissues (Burk *et al.*, 1995). Selenium is a multifunctional element that mainly serves as an effective antioxidant. The level of Se in animal body is closely pertinent to their antioxidant capacity and resistance to all kinds of diseases.

Alleviation of negative effects of oxidative stress on performance and nutrient digestibility

In our study, the effects of dietary selenium supplementation on performance and nutrient digestibiligy in stressed piglets were evaluated. Our results showed that the average body weight gain (ADG) of stressed piglets was increased by 9.93%, 16.76% and 20.37% after supplementation of the Se at the dosage of 0.2, 0.4 and 0.6 mg/kg diet, respectively. The ratio of feed intake to gain (F/G) was reduced by 10.16%, 12.30% and 13.90%, respectively. It's worthwhile to notice that the ADG and F/G of stressed piglets were close to those of control piglets when the dosage of 0.6 mg/kg was used. But the stressed piglets had a higher average daily feed intake (ADFI). Our results also showed that the apparent digestibility of CP, DM and GE has increased with the rising of Se concentration in the diet. All the results suggested that dietary supplementation of Se can alleviate the negative effects of oxidative stress on weaned piglets.

Enhancement of the activity of anti-oxidative enzymes

Previous studies showed that the synthesis of anti-oxidative enzymes such as GPx, SOD, and CAT significantly increased with the accumulating of ROS in animal body. The generation and scavenging of ROS keeps a physiological balance only if the animals are provided enough antioxidant nutrients. However, many abnormalities such as the stress or diseases disturbed the balance and induced an over-accumulating of ROS, which subsequently caused the damage of nucleic acids, proteins and cell membranes (Fang, 1997). In that case, more nutrients are needed for animals to alleviate or prevent the reduction of body antioxidant enzymes. In our study, we found that the serum activities of many antioxidant enzymes such as GSH, GPx, SOD and CAT tended to be enhanced with the increasing of dietary Se content.

Decrease of the forming of lipid hyperoxides and maintaining of the integrity of cell morphology and functions

Previous studies found that the PUFA is more sensitive than other substances to ROS. Thus, the forming of lipid hyperoxides has been normally considered as the mark of oxidative damage (Benzie, 1996).

Currently, the MDA content is the most widely used parameter to determine the quantity of lipid hyperoxides. Our results showed that both the liver and serum MDA content significantly increased in oxidative-stressed piglets ($P<0.05$). However, the serum MDA concentration acutely decreased when piglets received a diet supplemented with Se ($P<0.01$). We also found a linear correlation between the serum MDA and dietary Se levels. Usually, the higher dietary Se level has resulted in a lower serum MDA concentration. Therefore, the dietary Se supplementation is capable of maintaining the integrity of cell morphology and functions by preventing membranes from oxidative damage.

Decrease of incidences of DNA damage and cell apoptosis

Usually, the cell apoptosis is characterized by the activation of endonuclease and the forming of large numbers of DNA fragments. The results from agarose gel electrophoresis showed a marked DNA ladder. Our results showed that the DNA fragmentation of liver and spleen cells occurred in oxidative-stressed piglets. Other pathological changes including nuclear membrane expansion, shrinkage, nuclear fragmentation, chromatin condensation, and mitochondrial swelling were both observed. Compared with the piglets receiving a diet containing 0.2 mg/kg Se, there were only few pathological changes observed in piglets receiving a diet containing 0.4 mg/kg Se (all piglets received the diquat injection). Thus, dietary supplementation of Se is an effective approach to prevent the cell apoptosis.

4. The anti-oxidative mechanisms of selenium

Regulation by the hypothalamus-pituitary-thyroid axis

The complicated network of nerve-endocrine-immunity plays a crucial role in physiological events. The influences of oxidative stress on each link of the network will directly or indirectly affect the immune system. The thyroid is the biggest endocrine gland in animals. The secretion of thyroxin T_4 and T_3 are mainly regulated by the hypothalamus-pituitary-thyroid axis. However, the redox status of thyroid cells may also affect its secretion.

Our results showed that the concentrations of blood T_3, T_4 and the ratio of T_3 to T_4 ($P<0.05$) significantly elevated ($P<0.05$) in healthy piglets receiving a Se- containing diet. We obtained the same results in oxidative-stressed piglets with the exclusion of T_4 which tended to be decreased. Eder (2000) showed that both the total and dissociative T_4, and the ratio of T_4/T_3 has increased 1.32, 1.87 and 1.24 fold, respectively when the weaned piglets receiving a diet containing 15% oxidized sunflower oil (Eder and Stangl, 2000). This is probably due to the consuming of Se that has been usually recognized as the most important antioxidant nutrient. It is well known that Se plays an important role in activating the conversion of thyroxin from T_3 to T_4. Thus, the deficiency in Se declines the activity of iodothyronine deiodinase and subsequently increases the ratio of T_4/T_3. However, our results proved that dietary Se supplementation is capable of reversing such responses.

The thyroid-stimulating hormone (TSH) is considered to be the major pituitary hormone responsible the release of thyroid hormone. In our study, we found that the signal molecules of oxidative stress can promote the secreting of T_3 through activating the release of TSH via hypothalamus-pituitary-thyroid axis. The increased T_3 is capable of promoting the metabolic breakdown of nutrients, and eventually improves the capacity to resist the stress. Our results showed that the concentration of blood TSH significantly decreased in oxidative-stressed piglets. However, such enhancement can be restrained by dietary Se supplementation. The concentration of blood TSH tended to decline as the dietary Se content increased. Eder (2000) reported that the elevated serum TSH (caused by oxidated dietary fat) may result from the lipid hyperoxides that have been considered to be able to affect the thyroid axis (Eder and Stangl, 2000). However, the decreased serum TSH after Se supplementation could be attributed to the protective effects of Se itself on oxidative stress and the negative feedback of T_3.

Regulation by the nerve-endocrine-immune network

The nerve, endocrine and immune system forms a broad network that functions in physiolgical regulation. The influences of oxidative stress on each link of the network will directly or indirectly affect the immune system. A large number of studies have found that the oxidative stress pomotes the secreting of many inflammatory cytokines such as acute phase proteins (CRP), TNF-α, IL-1β, IL-6, IL-8 and adhesion melecules. The elevated levels of cytokines indicated the presence of the systemic inflammation in the body (Zhang et al., 2005). On the other hand, these cytokines can be used as the peripheral signals for systemic stress (LeMay et al., 1990). Our results showed that both the serum IL-1β and IL-6 significantly increased in stressed piglets. However, the abnormal increases in serum IL-1β and IL-6 can be suppressed by dietary Se supplementation. Many researchers believed that dietary supplementation of VC and VE may be effective to reduce the secreting of IL-β by monocytes which is induced by immunosuppression (Cannon, 1991). In our study, we found that the activation of cytokines in stressed piglets has been restrained by dietary Se supplementation. These results possibly indicated that the secretion of inflammatory cytokines is helpful for the defense system. However, an excessive secretion may induce some grave diseases that is harmful or fatal for animals (Meer, 1989). In our study, the dietary Se supplementation is capable of reducing the level of cytokines, which suggests that Se may prevent animal body from oxidative damage through the regulations of cytokine secretion via the nerve-endocrine-immune regulating network.

The molecular regulating mechanisms of oxidative stress

The nuclear factor kB (NF-kB) is a nuclear transcription factor that regulates expression of a large number of genes related to regulation of apoptosis, inflammation and various autoimmune diseases. The activation of NF-kB is thought to be a part of the stress response as it is activated by a variety of stimuli that include growth factors, cytokines, and oxidative stress. In our study, we found that oxidative stress significantly elevated the expression of NF-kB, NLS p65 and P-p65 in liver and spleen. However, the expression level of these proteins significantly decreased after dietary Se supplementation. Previous study indicated that the level of intracelluar ROS has been elevated by stimulatings such as TNF, IL-1 and UV and ionizing radiations. And the activity of IkapaB kinase complex (IKK) significantly increased within 2 mins after stimulating. They also found that H_2O_2 is an effective inducer for NF-kB activation. However, no matter dealing the cultured cells with Se compounds and the over-expression of both cyGPX and phospholipin peroxidase in cytoplasm, both of them were capable of declining the activity of NF-kB (Wang *et al.*, 2002). Therefore, the Se may exert its antioxidant role through NF-kB pathway (Figure 1).

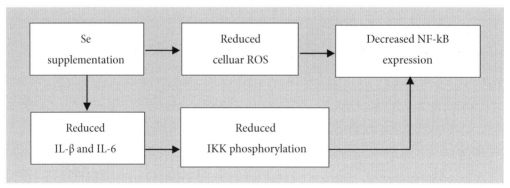

Figure 1. The regulations of NF-kB by Se.

5. Conclusion

With the advances of radical biology in the past decades, its basic theory has been fused into the field of life sciences including the agricultural science. Being one of the important parts of this field, the study of radical biology in animal science mainly focused on experimental animals such as the mice and rat. However, few of researches concern the livestock and poultry. Our studies indicated that oxidative stress not only greatly harms the health and growth performance of weaned piglets, but alters the requirments for some antioxidant nutrients. However, a rational supply of nutrients for animals significantly alleviates the detrimental effects that is resulted from the oxidative stress. Therefore, further in-depth studies about the underlying mechanisms and the nutritional control guidelines for oxidative stress are helpful to accelerate the realization of the disease-resistant productions for domestic animals and poultry.

Acknowledgement

This review is mainly based on researches in our lab. Most results have not published before. The research was granted by Program for Changjiang Scholars and Innovative Research Team in University with grant No.IRTO555-5, China Ministry of Education

References

Benzie, I.F.F., 1996. Lipid peroxidation:a review of causes,consequences, measurement dietary influences. Int.J.Food Sci.Nutr, 47: 233-261.

Bird, A.R.., Croom, W.J., Taylor, I.L., 1996. Peptide Regulation of Intestinal Glucose Absorption1. Animal. Science, 74: 2523-2540.

Burk, R.F., Hill, K.E., Awad, J.A., 1995. Pathogenesis of diquat induced liver necrosis in selenium deficient rats:assessment of the roles of lipid peroxidation and selenoprotein P[J]. Hepatology, 21: 561-569.

Cannon, J.G., 1991. Acute phase response in exercise.II.Association between vitamin E,cytokines and muscle proteolysis. Am J Physiol, 260: 1235-1240.

Cheng, Wei-li, Ding, Ming-min, He, Qiang, et al., 2005. The hemocyte parameters analysis of hypertension-coronary artery disease and diabetes mellitus patient. China Resting Therapy Medicine, 14: 229-230. (in Chinese).

Eder, K., Keller, U., Hirche, F., Bransch, C., 2003. Thermally oxidized dietary fats increase the susceptibility of rat LDL to lipid peroxidation but not their uptake by macrophages. J.Nutr, 133: 2830-2837.

Eder, K., Stangl, G.I., 2000. Plasma thyroxine and cholesterol concentrations of minizture pigs are influenced by thermally oxidized dietary lipids. J. Nutr, 130: 116-121.

Fang, Yun-zhong, 1997. The initial suggestion of application free radical biology and medicine. Advance of Free Radical Life Science, 1-3. (in Chinese).

Fang, Yun-zhong, Yang Sheng, Wu, Guo-yang, 2004. Free radical steady dynamic state. Advance of Physio-science, 35: 199-204. (in Chinese).

Fang, Yun-zhong, Yang, Sheng, Wu, Guo-yang, 2003. The relationship of free radical, anti-oxidant, nutriment and health. J Nutr, 25: 337-343. (in Chinese).

Farrington, J.A., Ebert, M., Land, E.J., Fletcher, K., 1973. Bipyridylium quaternary salts and related compounds. V. Pulse radiolysis studies of the reaction of paraquat radical with oxygen.Implication for the mode of action of bipyridyl herbicides. Biochim. Biophys. Acta, 314: 372-381.

Fu, Y., Cheng, W.H., Porres, J.M., Ross, D.A., Lei, X.G., 1999. Knockout of cellular glutathione peroxidase gene renders mice susceptible to diquat-induced oxidative stress. Free Radic.Biol.Med, 27: 605-611.

Galan, P., Preziosi, P., Monget, A.L., 1997. Effects of trace elements and/or vitamin supplementation on vitamin and mineral status,free radical metabolism and immunological markers in elderly long hospitalized subjects. Int J Vit Nutr Res, 67: 450-460.

Ghishan, F.K., Arab, N., Bulus, N., Said, H., Pietsch, J., Abumrad, N., 1990. Glutamine transport by human intestinal basolateral membrane vesicle1. American Society for Clinical Nutrition, 51: 612-616.

Guan, Qiu-hua, Liu, De-hui, 2000. Gucose transport protein and diabetes mellitus. Journal of China experimental zoology, 10: 175-180. (in Chinese).

Han, H.J., Choi, H.J., Park, S.H., 2000. High glucose-induced inhibition of -methyl-D-glucopyranoside uptake is mediated by protein kinase C-dependent activation of arachidonic acid release in primary cultured rabbit renal proximal tubule cells. J Cell Physiol, 183: 355-363.

Heuil, A.J., Meddings, J.B., 2001. Oxidative and drug-induced alterations in brush border membrane hemileaflet fluidity, functional consequences for glucose transport. Biochimica et Biophysica Acta (BBA) - Biomembranes, 1510: 342-353.

Huang, Ching-Jang, Cheung, N.S., Lu, V.R., 1988. Effects of deteriorated frying oil and dietary protein levels on liver on liver microsomal enzymes in rats. J.Am.Oil.Chem. Soc, 65: 1796-1803.

Kerr, J.E.R., Harmon, B.V., 1991. Definition and incidence of apoptosis:an historical perspective. In: Tomei, L.D., Cope, F.O. (eds.) Apoptosis:the molecular basis of cell death. Cold Spring Harbor Laboratory Press, Plainview, NY, USA, p. 5-29.

LeMay, L.G., Vander, A.J., Kluger, M.J., 1990. The effects of psychological stress on plasma interleukin-6 activity in rats. Physiol Behav, 47: 957-961.

Li, Li-juan, Chen, Dai-wen, Yu, Bing, 2007. Effect of oxidative stress on immune function of weanling pigs. J Anim Nutr. (in Chinese).

Liu, Song-yan, Li, Da-peng, Zhuang, Ping, et al., 2006 The research advancement of oxidative stress biomarker of aquatic animal. Water Conservancy Related Fisheries, 26: 16-19. (in Chinese).

Marchionatti, A., Alisio, A., Diaz de Barboza, G., Baudino, V., Tolosa de Talamoni, N., 2001. DL-Buthionine-S,R-sulfoximine affects intestinal alkaline phosphatase activity. Comparative Biochemistry and Physiology Part C: Toxicology & Pharmacology, 129: 85-91.

Meer, V.M., 1989. Options for the treatment of serious infections with interleukin-1. Biotherapy, 1: 313-317.

Prabhu, R., Anup, R., Balasubramanian, K.A., 2000. Surgical Stress Induces Phospholipid Degradation in the Intestinal Brush Border Membrane. Journal of Surgical Research, 94: 178-184.

Schonbaum, G.R., Chance, C., 1976. Catalase. In: Boyer, P.D. (ed.) The Enzvs (2nd Ed.). Academic Press, New York, USA, p. 363-468.

Skufca, P., Brandsch, C., Hirche, F., 2003. Effects of a dietary thermally oxidized fat on thyroid morphology and mRNA concentrations of thyroidal iodide transporter and thyroid peroxidase in rats. Annals of nutrition - Metabolism, 47: 207-213.

Tarpey, P.S., Wood, I.S., Shirazi-Beechey, S.P., Beechey, B., 1995. Amino acid sequence and the cellular location of the Na(+)-dependent D-glucose symporters (SGLT1) in the ovine enterocyte and the parotid acinar cell. Biochem J 312: 293-300.

Tong, Yi-qiu, 2003. Technology standard specification and data readout practical handbook of hospital health examination chemical analysis. Jilin AV Press, Jilin , China. (in Chinese).

Wang, Di-xun, Jin, Hui-ming, 2002. Human body pathophysiology. Beijing People Health Press, Beijing, China, p. 515-516. (in chinese).

Wang, Hai-rong, Li, Jian-jun, Jiang, Xi-jia, et al., 2002. NF-kB and oxidical dress. Advance of Angiocardiopathy Science, 23: 166-170. (in Chinese).

Wu, Yong-kui, Hu, Zhong-ming, 2005. Animal stress medicine and the molecule regulation mechanism of the dress[J]. Journal of China Veterinary, 25:557-560. (in Chinese).

Xiong, Li-fan, Li, Shu-ren, 2003. Clinical Laboratory Basis. People Health Press, Beijing, China. (in Chinese).

Yuan, Shi-bin, Chen, Dai-wen, Zhang, Ke-ying et al., 2007. Effects of oxidative stress on growth performance, nutrient digestibilities and activities of antioxidant enzymes of weanling pigs. Asia-Austr. J. Anim. Nutr. 20: 1600-1605.

Zhang, Ming-zhu, He, Guang-yuan, Zhu, Hong, 2005. The viestigation of leucocyte parameters change before or after chemotherapy malignancy patient. Present Tumor Medicine, 13: 334-335. (in Chinese).

Zhao, yin, Wei, Jun, Wang, Zhi-Wei, et al., 2001. The clinical significance of RBC-MC and thrombocyte parameters change of cerbral infarction patient. Journal of Clinical Examination, 19: 236-236. (in Chinese).

Functional amino acids in swine nutrition and production

Guoyao Wu[1], Fuller W. Bazer[1], Robert C. Burghardt[2], Gregory A. Johnson[2], Sung Woo Kim[3], Darrell A. Knabe[1], Xilong Li[1], M. Carey Satterfield[1], Stephen B. Smith[1] and Thomas E. Spencer[1]

[1]Department of Animal Science, Texas A&M University, College Station, TX 77843, USA; g-wu@tamu.edu

[2]Department of Veterinary Integrative Bioscience, Texas A&M University, College Station, TX 77843, USA

[3]Department of Animal Science, North Carolina State University, Raleigh, NC 27695, USA

Abstract

Amino acids were traditionally classified as nutritionally essential or nonessential for swine based on nitrogen balance and growth. It was also assumed without much evidence that pigs could synthesize sufficient amounts of all nonessential amino acids to support maximum production performance. Thus, over the past 50 years, much emphasis has been placed on the roles for dietary essential amino acids as building blocks for tissue proteins. Disappointingly, the current version of NRC does not recommend dietary requirements of so called 'nonessential amino acids' by neonatal, post-weaning, growing-finishing, or gestating pigs. However, a large body of literature shows that these nonessential amino acids, particularly glutamine and arginine, play important roles in regulating gene expression at both transcriptional and translational levels in animals. Additionally, both isotopic and digestive studies have established that large amounts of amino acids in the enteral diet are degraded by the small intestine during the first pass. Thus, only 5% of glutamate and aspartate, 30-33% of glutamine, and 60-65% of proline and arginine in the diet enter the portal circulation. Dynamic synthesis of amino acids via inter-organ metabolism depends on essential amino acids and may be suboptimal in pigs at various stages of the life cycle. Amino acids participate in cell signaling via mammalian target of rapamycin, AMP-activated protein kinase, extracellular signal-related kinase, Jun kinase, mitogen-activated protein kinase, and gases (NO, CO and H_2S). Exquisite integration of these regulatory networks has profound effects on cell proliferation, differentiation, metabolism, homeostasis, survival, and function. Importantly, recent advances in understanding of functional amino acids are transforming the practice of swine nutrition. Particularly, dietary supplementation with 1% L-glutamine or L-arginine prevented intestinal dysfunction (a significant problem in swine production) in low-birth-weight piglets and early-weaned pigs, while increasing their growth performance and survival. Also, dietary supplementation with 1% L-arginine improved spermatogenesis and sperm quality in boars, growth and survival of milk-fed and weaned piglets, muscle gain and meat quality in finishing pigs, as well as litter size and fetal growth in gilts. Availability of feed-grade amino acids (e.g., arginine, glutamine, leucine, and proline) is expected to improve the efficiency and quality of pork production worldwide.

1. Introduction

Based on nitrogen balance and growth, amino acids have traditionally been classified as nutritionally essential or nonessential for animals (Baker, 2005; Kim *et al.*, 2005). The carbon skeletons of essential amino acids are not synthesized by animal cells and, therefore, must be provided in the diet. In contrast, inter-organ metabolism of amino acids results in the *de novo* synthesis of nonessential amino acids (Bergen and Wu, 2009; Wu *et al.*, 2009). Growing evidence shows that pigs cannot synthesize sufficient amounts of all nonessential amino acids to support maximum production performance (Wu *et al.*, 1996a; Kim and Wu, 2004; Mateo *et al.*, 2008). Clearly, the classic definition of essential and nonessential amino acids has major conceptual limitations, because the functions of amino acids beyond protein synthesis have not been taken into consideration (Wu *et al.*, 2005; Wu, 2009). Disappointingly, the National Research Council (NRC) does not recommend dietary requirements of nonessential amino acids by neonatal, post-weaning, growing, or gestating swine (NRC, 1998).

A growing body of evidence supports the notion that amino acids play versatile roles in physiology and nutrition (Reeds and Burrin, 2001; Jobgen *et al.*, 2006). Results of recent studies have led to development of the concept of functional amino acids, which are defined as amino acids that regulate key metabolic pathways to benefit health, survival, growth, development, and reproduction of animals (Wu, 2009). Functional amino acids (e.g., arginine, cysteine, glutamine, leucine, and proline) can substantially improve the efficiency of utilization of dietary proteins in pigs (Kim and Wu, 2004; Li *et al.*, 2007; Wang *et al.*, 2008a). The objective of this article is to highlight recent advances in understanding of the roles for functional amino acids in whole-body nitrogen metabolism, gene expression, cell signaling, and dynamic swine nutrition.

2. Efficiency of the utilization of dietary amino acids

Synthesis of amino acids

Milk has traditionally been thought to provide adequate amounts of all amino acids to neonates (NRC, 1998). However, this notion is not consistent with the recent finding that sow milk is markedly deficient in arginine and that this deficiency is a major factor limiting maximum growth of 7- to 21-day-old suckling piglets (Wu and Knabe, 1994; Wu et al., 2004). Considering the combined use of arginine (0.37 g/kg body weight) via arginase, nitric oxide synthase, arginine decarboxylase, and arginine:glycine amidinotransferase pathways (Wu and Morris, 1998), milk protein provides at most 40% of arginine for protein deposition in the 14-day-old pig (Table 1) as previously reported for the 7-day-old pig (Wu et al., 2004). Besides arginine, the amount of dietary proline that enters the portal vein is inadequate to support proline requirements for protein synthesis in the piglet (Table 1). Based on a degradation rate (0.93 g/kg body weight per day) of intravenously infused proline in young pigs (Murphy et al., 1996), the de novo synthesis of proline must occur at a rate of at least 1.11 g/kg body weight per day (or at least 60% of the proline need for protein accretion). The dynamic metabolism of proline is necessary for maintaining the homeostasis of proline and arginine in animals. Additionally, based on glycine content of sow milk and the body, milk-derived glycine meets less than 30% of the need for protein synthesis in the piglet (Table 1).

Collagen represents 30% of total protein in young pigs (Wu et al., 1999). Because the conversion of proline into hydroxyproline in protein is a posttranslational event, the amount of proline plus hydroxyproline reflects the proline requirement for protein synthesis. Except for arginine and proline, other essential amino acids in sow's milk are sufficient to support protein synthesis in suckling piglets. For example, in a 14-day-old pig gaining 235 g/day, 40-45% of these essential amino acids from the diet are deposited as tissue proteins in the body (Table 1). Branched-chain amino acids account for one-third of total essential amino acids absorbed from the small intestine. Approximately 60 to 90% of the dietary essential amino acids that enter the portal vein are utilized for protein synthesis in suckling piglets, with the lowest rate (62%) for lysine and the highest rate (89%) for valine (Table 1). These results indicate that protein synthesis is the major pathway for utilization of dietary essential amino acids in neonates, with the remainder being used for the production of nonessential amino acids and other nitrogenous products (Figure 1).

Aspartate plus asparagine and glutamate plus glutamine represent 23% and 42%, respectively, of total nonessential amino acids in sow milk (Table 1). However, almost none of the dietary aspartate and

Table 1. Composition of amino acids in sow milk (protein-bound plus free) and efficiency of milk protein for growth in the 14-day-old pig (3.9 kg body weight)[1].

AA	AA in sow milk[2] (g/l of whole milk)	AA intake by the pig[3] (g/day)	AA entering the portal vein[4] (g/day)
EAA	27.3	24.9	18.0
Arginine	1.43	1.31	1.06
Histidine	0.92	0.84	0.76
Isoleucine	2.28	2.08	1.41
Leucine	4.46	4.07	2.78
Lysine	4.08	3.73	3.09
Methionine	1.04	0.95	0.85
Phenylalanine	2.03	1.85	1.50
Proline	5.59	5.10	3.12
Hydroxyproline	0.00	0.00	0.00
Pro + OH-Pro	5.59	5.10	3.12
Threonine	2.29	2.09	1.32
Tryptophan	0.66	0.60	0.52
Valine	2.54	2.32	1.51
NEAA	22.7	19.5	9.59
Alanine	1.97	1.80	1.38
Asparagine	2.53	2.31	2.00
Aspartate	2.59	2.36	0.11
Cysteine	0.72	0.66	0.50
Glutamate	4.57	4.17	0.21
Glutamine	4.87	4.45	1.42
Glycine	1.12	1.02	0.82
Serine	2.35	2.15	1.72
Tyrosine	1.94	1.77	1.43
Total AA	50.0	44.8	27.6

[1]The molecular weights of intact amino acids were used for all the calculations.
[2]Milk was obtained on Days 7-21 of lactation for analysis of total amino acids (Kim and Wu, 2004).
[3]Milk consumption of 913±35 ml/day (mean ± SEM, n=30), using the weigh-suckle-weigh technique (*Wu et al.* 2000).
[4]Calculated on the basis of (1) the ileal digestibility (%) of amino acids (Arg, 90; His, 100; Ile, 90; Leu, 91; Lys, 92; Met, 99; Phe, 90; Pro, 94; Thr, 84; Trp, 96; Val, 87; Ala, 85; Asn, 96; Asp, 96; Cys, 84; Glu, 100; Gln, 100; Gly, 89; Ser, 89; Tyr, 90) (Mavromichalis *et al.*, 2001); and (2) the bioavailability (%) of orally administered amino acids (Arg, 90; His, 90; Ile, 75; Leu, 75; Lys, 90; Met, 90; Phe, 90; Pro, 65.; Thr, 75; Trp, 90; Val, 75; Ala, 90; Asn, 90; Asp, 5; Cys, 90; Glu, 5; Gln, 32; Gly, 90; Ser, 90; Tyr, 90). Products of intestinal AA metabolism that enter the portal vein are not included.

AA content in the pig[5]	AA accretion in the pig[6]	AA accretion in the pig/ AA intake from the diet[7]
(mg/g wet weight)	(g/day)	(%)
69.3±0.27	16.4	65.9 (91.1)
9.13±0.06	2.15	164 (203)
2.79±0.05	0.66	78.6 (86.8)
4.75±0.06	1.12	53.8 (79.4)
9.20 ±0.08	2.16	53.1 (77.7)
8.11±0.07	1.91	51.2 (61.8)
2.54±0.04	0.60	63.2 (70.6)
4.67±0.07	1.10	59.5 (73.3)
11.3±0.08	2.66	---
4.96±0.06	1.17	---
16.3±0.10	3.83	75.1 (123)
4.72±0.07	1.11	53.1 (84.1)
1.47±0.03	0.35	58.3 (67.3)
5.71±0.07	1.34	57.8 (88.7)
63.0±0.31	14.8	75.9 (154)
8.85±0.08	2.08	116 (151)
4.76±0.07	1.12	48.5 (56.0)
5.74±0.10	1.35	57.2 (1,227)
1.77±0.04	0.42	63.6 (84.0)
10.5±0.12	2.47	59.2 (1,176)
6.78±0.06	1.59	35.7 (112)
15.2±0.13	3.57	350 (435)
5.97±0.09	1.40	65.1 (81.4)
3.56±0.05	0.84	47.5 (58.7)
132±0.45	31.2	69.6

[5]Amino acid composition was determined as described by Wu *et al.* (1999). Values are means±SEM, n=8.

[6]Calculated on the basis of a body-weight gain of 235 g/day.

[7]Ratios of AA accretion in the pig to AA entering the portal vein from the small intestine are given in parentheses.

AA, amino acids; EAA, Nutritionally essential amino acids; NEAA, Nutritionally nonessential amino acids; OH-Pro, hydroxyproline; Pro, proline.

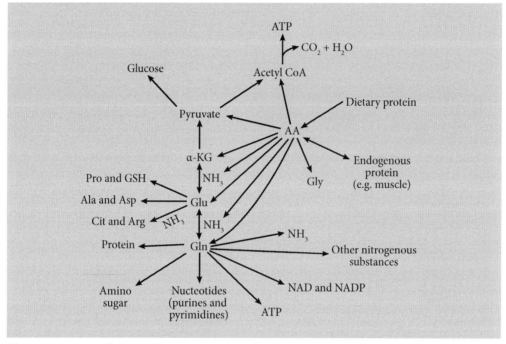

Figure 1. An overall view of the degradation of amino acids in pigs. Degradation of essential amino acids via inter-organ cooperation results in synthesis of nonessential amino acids. Glutamine plays a central role in synthesis of glutamate, ammonia, citrulline, arginine, proline, alanine, aspartate, glutathione, aminosugars, nucleotides, NAD, NADP, ATP, and glucose. Abbreviations: AA, amino acids; Ala, alanine; Arg, arginine; Asp, aspartate; Cit, citrulline; Glu, glutamate; Gln, glutamine; Gly, glycine; GSH, glutathione; α-KG, α-ketoglutarate; Pro, proline.

glutamate enters the portal circulation (Table 1), resulting in low concentrations of these amino acids in plasma of the pig (Wu *et al.*, 1996b). Among nonessential amino acids, only asparagine, cysteine, serine, and tyrosine in milk are sufficient to meet requirements for protein accretion. In contrast, the milk provides at most 66%, 8%, 9%, and 23% of alanine, aspartate, glutamate, and glycine for protein deposition in suckling pigs, respectively. Considering extensive utilization of arterial glutamine by both enterocytes and other cell types (including kidneys and lymphocytes) (Wu, 2009), dietary glutamine is also substantially inadequate for protein synthesis in extra-intestinal tissues of piglets. Because information about oxidation rates of all amino acids in swine is not available, it is currently not possible to estimate rates of their total synthesis *in vivo*. Nonetheless, there is a high rate of whole-body glutamine utilization in the piglet (at least 1.65 g/kg body weight per day). In young pigs, a large amount of glutamine is oxidized to CO_2 (0.47

g glutamine/kg body weight per day) (Stoll *et al.*, 1999) and deposited in body proteins (0.41 g glutamine/kg body weight per day) (Table 1). Given a much smaller amount of dietary glutamine entering the systemic circulation (0.36 g/kg body weight per day) (Table 1), the rate of *de novo* synthesis of glutamine is likely the greatest in the suckling piglet (at least 0.88 g glutamine/kg body weight per day). Also, based on protein accretion, net rates of synthesis should also be high for glycine, glutamate, aspartate, arginine, and alanine in the neonate (Table 1). Because the half lives of glutamate, aspartate, arginine, alanine, and glutamine in the blood of sow-reared piglets are short (ranging from 0.64 h for glutamine to 1.08 h for glutamate) (Table 2), endogenous synthesis is important for maintaining their homeostasis in the body (Flynn and Wu, 1996; Wu, 2009).

Both milk and typical corn- and soybean meal-based diets contain large amounts of branched-chain amino acids (27%), aspartate plus asparagine (11%), and glutamate plus glutamine (17%) (Tables 3 and 4). Ratios of essential amino acids to lysine in the diets are 10-15% higher than ratios of nonessential amino acids to lysine. In both suckling and weaned pigs, the entry of dietary amino acids from the small intestine into the portal vein is insufficient to meet requirements for protein synthesis. Thus, amounts of essential amino acids in the diet must exceed their requirements for protein accretion. Efficiency of dietary essential and nonessential amino acids for protein synthesis is 35% and 26% lower, respectively, in post-weaning pigs than in suckling piglets. This may be explained by: (1) greater rates of entry of dietary amino acids into the systemic circulation (Tables 1 and 3); (2) greater sensitivity of protein synthesis to positive regulators (e.g., insulin and growth hormone) in young than in older pigs (Suryawan *et al.*, 2009). In contrast to the common belief (NRC, 1998; Kim *et al.*, 2005), the plant protein-based diet does not provide adequate arginine, proline, aspartate, glutamate, glutamine, or glycine for young weaned pigs (Table 3). Major differences between the weaning diet and milk are that the weaning diet is not deficient in alanine and provides high ratios for arginine, histidine, phenylalanine, alanine, aspartate and glycine relative to lysine (Table 5).

Our extensive studies established that arginine is synthesized from glutamine, glutamate and proline (abundant amino acids in milk) in neonatal pigs, which has important nutritional and physiological significance (Wu and Knabe, 1995; Flynn and Wu, 1996; Wu, 1997). In contrast, little is known about how glycine is produced in the piglet. Daily synthesis of glycine should be at least 2.75 g in the 14-day-old pig (Table 1). Although biochemistry textbooks state that glycine is synthesized from serine, 81% of milk-born serine is utilized for protein deposition and the diet provides at most 0.32 g of serine for glycine formation (Table 1). Therefore, 90% of glycine must be synthesized from precursors other than serine. Currently, the underlying pathways

Table 2. Half-lives ($T_{1/2}$) of L-amino acids in blood and the bioavailability (F) of orally administered L-amino acids in pigs.

	Arginine	Aspartate	Citrulline	Glutamate	Glutamine	Leucine	Lysine	Proline
14-day-old pigs nursed by sows[1]								
$T_{1/2}$, h	0.66±0.04	1.06±0.06	1.27±0.08	1.08±0.08	0.64±0.03	1.16±0.09	1.24±0.10	0.92±0.05
F, %	90±7	5±0.4	91±6	5±0.6	32±3	75±6	90±8	65±5
35-day-old pigs weaned at 21 days of age[2]								
$T_{1/2}$, h	0.67±0.05	1.12±0.08	1.31±0.10	1.19±0.09	0.66±0.05	1.28±0.10	1.33±0.15	0.95±0.07
F, %	60±4	5±0.5	85±5	4±0.7	33±4	66±5	75±6	61±4
Pregnant gilts (105 days of gestation)[3]								
$T_{1/2}$, h	0.71±0.08	1.08±0.10	1.38±0.12	1.14±0.11	0.67±0.06	1.24±0.13	1.21±0.12	0.90±0.08
F, %	59±6	6±0.8	75±9	5±0.7	30±4	63±7	72±8	60±6

Values are means ± SEM, n=6. Pigs received an intravenous or oral bolus of each of the following L-amino acids (mg/kg body weight): arginine-HCl, 60; aspartate, 10; citrulline, 50; glutamate, 25; glutamine, 100; leucine, 50; lysine-HCl, 50; and proline, 50. Blood sampling, determination of the half lives of intravenously administered amino acids in blood, and calculation of the bioavailability of orally administered amino acids was performed, as described by Wu *et al.* (2007).

[1]Lactating sows were fed a corn- and soybean meal-based diet containing 18.7% crude protein (Mateo *et al.*, 2008).

[2]Pigs were fed a corn- and soybean meal-based diet containing 21.0% crude protein (Wang *et al.*, 2008a).

[3]Gestating gilts were fed a corn- and soybean meal-based diet containing 12.2% crude protein (Mateo *et al.*, 2007).

Table 3. Utilization of glutamine and glutamate in the enteral diet by weaned young pigs.[1]

Parameter	Arg	Asp	Asn	Glu	Gln	Leu	Lys	Pro
Content in diet (g/kg DM)	14.8	14.5	10.5	19.2	20.5	19.8	15.9	17.4
Intake from diet (mg/kg BW per day)	744	729	528	966	1,031	996	800	875
True ileal digestibility (%)[2]	90	86	85	87	84	88	84	86
Availability from diet (mg/kg BW per day)	670	627	449	840	866	880	672	753
Entering portal vein (mg/kg BW per day)	402	31	391	34	286	582	504	459
Utilization of dietary components by SI (mg/kg BW per day)	268	596	58	806	580	298	168	294
Flux from arterial plasma into SI (mg/kg BW per day)[3]	0	0	0	0	264	0	0	0
Utilization by SI (mg/kg BW per day)	268	596	58	806	844	298	168	294

[1] Values are means for ten 35-day-old pigs (8.7 kg; Yorkshire×Landrace dams and Duroc×Hampshire sires) weaned at 21 days of age to a corn- and soybean meal-based diet containing 21.5% crude protein) (Wang et al., 2008a). Pigs had free access to the diet and drinking water. Feed intake was 50.3 g DM/kg BW per day. Asparagine plus aspartate and glutamine plus glutamate in the Arg diet were analyzed by HPLC after acid hydrolysis (Wu et al., 1999). The ratios of asparagine:aspartate and glutamine:glutamate in the diet were determined using a bioassay method, which involved incubation of 50 mg of a finely ground sample or 50 mg water (blank) with 2.5 ml porcine gastric fluid (2h at 37 °C) and, after neutralization, with 5 ml of porcine small-intestinal luminal fluid (4h at 37 °C). Porcine gastric and small-intestinal luminal fluids were obtained from 12-h food-deprived 35-day-old pigs weaned at 21 days of age.

[2] True ileal digestibility was determined on the basis of the collection of ileal digesta samples from the pigs fed the corn- and soybean meal-based diet and a corn starch-based nitrogen-free diet (Kong et al., 2008).

[3] Following a 12-h period of food deprivation, the jejunal artery and jejunal vein of pigs were cannulated (Wu et al., 1994) and blood flow was measured by an indicator dilution technique involving infusion of p-aminohippuric acid (Yen and Killerfer, 1987). Amino acid uptake by the small intestine = Arterial-venous differences in concentrations×Rate of small-intestine blood flow×(1−Hematocrit). Values for arterial-venous differences in concentrations of glutamine in plasma, small-intestine blood flow, and hematocrit were 56.4±4.8 µmol/l plasma, 33.7±3.1 ml/min per kg body weight and 0.33±0.02 (mean ± SEM, n=6), respectively.

BW, body weight; SI, small intestine.

Table 4. Composition of amino acids in the weaning diet and efficiency of dietary protein for growth in the 30-day-old pig (7.8 kg body weight) weaned at 21 days of age[1].

AA	AA in the weanling diet[2] (g/kg)	AA intake by the pig[3] (g/day)	AA entering the portal vein from the diet[4] (g/day)
EAA	110	38.7	23.0
Arginine	13.2	4.64	2.50
Histidine	5.73	2.01	1.43
Isoleucine	8.91	3.13	1.79
Leucine	17.8	6.25	3.63
Lysine	14.2	4.98	3.14
Methionine	3.58	1.26	0.89
Phenylalanine	9.93	3.49	2.37
Proline	15.8	5.55	2.91
Hydroxyproline	0.00	0.00	0.00
Pro + OH-Pro	15.8	5.55	2.91
Threonine	8.52	2.99	1.79
Tryptophan	2.49	0.87	0.60
Valine	9.96	3.50	1.94
NEAA	99.2	34.9	15.2
Alanine	13.0	4.57	3.50
Asparagine	9.40	3.30	2.44
Aspartate	13.2	4.64	0.23
Cysteine	3.74	1.32	0.91
Glutamate	17.2	6.04	0.21
Glutamine	18.4	6.46	1.79
Glycine	8.81	3.10	2.21
Serine	7.86	2.76	2.04
Tyrosine	7.62	2.68	1.89
Total AA	209	73.6	38.2

[1]The molecular weights of intact amino acids were used for all the calculations.
[2]Corn- and soybean meal-based diet containing 21.5% crude protein (Wang *et al.*, 2008a). Dry matter content in the diet was 89.5%.
[3]Feed intake (as-fed basis) was 45.0±3.8 g/kg body weight per day (mean ± SEM, n=20).
[4]Calculated on the basis of (1) the true ileal digestibility (%) of amino acids (Arg, 90; His, 89; Ile, 88; Leu, 88; Lys, 84; Met, 89; Phe, 87; Pro, 86; Thr, 83; Trp, 84; Val, 85; Ala, 88; Asn, 85; Asp, 86; Cys, 83; Glu, 87; Gln, 84.0; Gly, 87; Ser, 89; Tyr, 88); (2) the bioavailability (%) of orally administered amino acids (Arg, 60; His, 80 Ile, 65; Leu, 66; Lys, 75; Met, 79; Phe, 78; Pro, 61; Thr, 72; Trp, 81; Val, 65; Ala, 87; Asn, 87; Asp, 5; Cys, 83; Glu, 4; Gln, 33; Gly, 82; Ser, 83; Tyr, 80). Products of intestinal AA metabolism that enter the portal vein are not included.

AA content in the pig[5]	AA accretion in the pig[6]	AA accretion in the pig/ AA intake from the diet[7]
(mg/g wet weight)	(g/day)	(%)
72.8±0.53	21.2	54.8 (92.2)
9.52±0.11	2.76	59.5 (110)
2.92±0.08	0.85	42.3 (62.5)
4.97±0.07	1.44	46.0 (80.4)
9.61±0.13	2.79	44.6 (76.9)
8.48±0.09	2.46	49.4 (78.3)
2.63±0.06	0.76	60.3 (85.4)
4.82±0.10	1.40	40.1 (59.1)
12.1±0.14	3.51	---
5.33±0.11	1.55	---
17.4±0.16	5.06	91.2 (174)
4.93±0.09	1.43	47.8 (80.0)
1.56±0.06	0.45	51.7 (75.0)
5.93±0.10	1.72	49.1 (88.7)
67.9±0.68	19.7	56.4 (130)
9.24±0.16	2.68	58.6 (76.6)
5.06±0.10	1.47	44.5 (60.2)
6.02±0.13	1.75	37.7 (761)
1.86±0.06	0.54	40.9 (59.3)
11.9±0.17	3.45	57.1 (1,643)
7.20±0.09	2.09	32.4 (117)
16.5±0.18	4.79	155 (217)
6.23±0.14	1.81	65.6 (88.7)
3.82±0.09	1.11	41.4 (58.7)
141±0.92	40.8	55.4

[5]Amino acid composition was determined as described by Wu *et al.* (1999). Values are means±SEM, n=8.
[6]Calculated on the basis of a body-weight gain of 290 g/day.
[7]Ratios of AA accretion in the pig to AA entering the portal vein from the small intestine are given in parentheses.
AA, amino acids; EAA, Nutritionally essential amino acids; NEAA, Nutritionally nonessential amino acids; OH-Pro, hydroxyproline; Pro, proline.

Table 5. Patterns of amino acids in sow milk, the weaning diet, entry from the small intestine lumen into the portal vein of fed pigs, the pig body, and plasma[1].

AA	Patterns of AA in diet		Patterns of AA entering the portal vein	
	Sow milk[2]	Weaning diet[3]	Sow milk[4]	Weaning diet[5]
EAA	669	775	583	733
Arginine	35	93	34	80
Histidine	23	40	25	46
Isoleucine	56	63	46	57
Leucine	109	125	90	116
Lysine	100	100	100	100
Methionine	25	25	28	28
Phenylalanine	50	70	49	76
Proline	137	111	101	93
Threonine	56	60	43	57
Tryptophan	16	18	17	19
Valine	62	70	49	62
NEAA	556	699	310	484
Alanine	48	92	45	112
Asparagine	62	66	65	78
Aspartate	63	93	4	7
Cysteine	18	26	16	29
Glutamate	112	121	7	7
Glutamine	119	130	46	57
Glycine	27	62	27	70
Serine	58	55	56	65
Tyrosine	48	54	46	60
Total AA	1,225	1,472	893	1,217

[1] Lysine is used as the reference value (100). The ratio of amino acid to lysine is expressed as percent (g/g ×100).
[2] Sow's milk (Kim and Wu, 2004). Lysine content in the diet was 1.66% (dry matter basis).
[3] Corn- and soybean meal-based diet containing 21.5% crude protein (Wang *et al.*, 2008a). Lysine content in the diet was 1.59% (dry matter basis).
[4] Sow-reared 14-day-old pigs.
[5] Thirty-day-old pig weaned at 21 days of age to a corn- and soybean meal-based diet containing 21.5% crude protein (Wang *et al.*, 2008a).

Patterns of AA in the pig plasma		Patterns of AA in the pig body	
Sow milk[4]	Weaning diet[5]	Sow milk[6]	Weaning diet[7]
833	982	859	796
73	120	113	112
44	60	34	34
47	55	59	59
71	97	113	113
100	100	100	100
36	33	31	31
46	50	58	57
198	212	201	205
91	104	58	58
25	30	18	18
104	121	70	70
973	1,106	777	801
203	194	109	109
39	42	59	60
7	7	71	71
59	69	22	22
67	86	130	140
228	290	84	85
203	236	187	195
78	85	74	74
90	98	44	45
1,807	2,088	1,628	1,663

[6]Suckling 14-day-old pigs. Blood samples (1 ml) were obtained from the jugular vein at 1.5 h after feeding. Lysine concentration in plasma was 230 µM.

[7]Thirty-day-old pig weaned at 21 days of age to a corn- and soybean meal-based diet containing 21.5% crude protein (Wang et al., 2008a). Blood samples (1 ml) were obtained from the jugular vein at 1.5 h after feeding. Lysine concentration in plasma was 186 µM.

AA, amino acids; EAA, Nutritionally essential amino acids; NEAA, Nutritionally nonessential amino acids.

(including substrates and reactions) for glycine synthesis in pigs are not known.

Degradation of amino acids

It has been a long-standing belief that dietary amino acids enter the portal vein intact (Wu, 1998). However, this concept has recently been challenged by findings from studies with young pigs that both essential and nonessential amino acids in the enteral diet are degraded extensively by the small intestine in the first pass (Stoll and Burrin, 2006). Results of our studies indicate that nearly all of glutamate and aspartate, 67-70% of glutamine, and 30-40% of proline in the enteral diet are catabolized by the small intestine of neonatal, weaned, and gestating swine (Table 3). Similar results were obtained by Reeds and coworkers for glutamate and aspartate (Reeds *et al.*, 1996, 1997). Thus, only 5% of glutamate and aspartate, 30-33% of glutamine, as well as 60-65% of proline and arginine in the enteral diet enter the portal circulation. The rate of degradation of glutamate in the enteral diet by the small intestine is the greatest among amino acids, followed by glutamine, aspartate, and proline. Interestingly, the small intestine takes up a large amount of glutamine, but no glutamate or aspartate, from the arterial blood for catabolism (Table 3). Consequently, the rate of utilization of glutamine by the gut is greater than that for dietary glutamate. It is now known that bacteria in the lumen of the small intestine can degrade amino acids (including those from the enteral diet (Bergen and Wu, 2009). Furthermore, absorptive epithelial cells of the neonatal gut (enterocytes) extensively catabolize glutamate, aspartate, glutamine and proline (Wu an Knabe, 1995; Wu *et al.*, 1996c; Wu, 1997).

Nitrogenous products of glutamate and glutamine include ornithine, citrulline, arginine, aspartate, and alanine (Wu, 1998). Notably, there is little degradation of arginine by enterocytes of preweaning pigs (Wu *et al.*, 1996), whereas 40% of dietary arginine is utilized by the small intestine of post-weaning pigs in the first pass (Wu *et al.*, 2007). Some evidence shows that 25% to 60% of essential amino acids in the diet (e.g., 40% leucine, 30% isoleucine, 40% valine, 50% lysine, 50% methionine, 45% phenylalanine, and 60% threonine) may be extracted by the small intestine of weaned piglets in first pass (Stoll and Burrin, 2006). Using bolus oral or intravenous administration of the same dose of a test amino acid, we obtained lower values of first-pass intestinal use for histidine, lysine, methionine, phenylalanine, threonine, and tryptophan in 14-day-old suckling pigs (Table 1) and 35-day-old weaned pigs (Table 3). Less than 20% of the extracted amino acids are utilized for intestinal mucosal protein synthesis (Stoll and Burrin, 2006), and greater than 80% of the extracted amino acids are presumably degraded by enterocytes and/or microorganisms in the intestinal lumen.

Enterocytes of both pre- and post-weaning pigs have a high activity of branched-chain amino acid transaminase (BCAT) and, therefore, can extensively transaminate leucine, isoleucine and valine, with their α-ketoacids being partially decarboxylated to CO_2 (Chen *et al.*, 2007; 2009). However, these cells do not express key enzymes for degrading histidine, lysine, methionine, phenylalanine, threonine or tryptophan and have little ability to oxidize them to CO_2. The findings raised the possibility that the extensive catabolism of dietary essential amino acids by the pig small intestine may result from the action of lumenal microbes. Bacterial degradation of amino acids in the lumen of the gut contributes to the extensive recycling of nitrogen in the digestive tract (Table 6). It is likely that previous isotopic studies overestimated rates of extraction of essential amino acids in the diet by the pig small intestine. In support of this view, we found that, based on protein intake from diets and protein accretion in the body, the efficiency of conversion of dietary amino acids into body protein is 70% and 55%, respectively, for 14-day-old sow-reared pigs and 30-day-old pigs weaned at 21 days of age to a corn- and soybean meal-based diet (Table 3). Considering (1) ileal digestibilities of dietary amino acids; (2) extensive catabolism of dietary glutamate, glutamine, aspartate, and proline; and (3) oxidation of amino acids (at least 5-10% of absorbed amino acids) by extra-intestinal tissues, less than 20% and 35% of dietary essential amino acids would

Table 6. The sources and absorption of nitrogen in the lumen of the intestine.[1]

Source	Amount of N (mg/kg body wt)
Sources of N in the small intestine	
Diet	1,530
Endogenous source	500
Saliva	14
Stomach	180
Bile	50
Pancreatic secretions	56
Small-intestinal secretions	100
Microorganisms in the lumen	100
Absorption of N in the small intestine	1,600
Secretions of N into the lumen of large intestine	65
Absorption of N in the large intestine	320
Fecal N excretion	175

[1] Estimated from the 30-50 kg growing pig consuming an 18%-crude protein diet. Data are adapted from Fuller and Reeds (1998) and Bergen and Wu (2009). Dry matter intake is 5.3% of body weight per day.

be degraded during the first pass by the small intestine of 14-day-old suckling pigs and 30-day-old weaned pigs, respectively.

Amino acid metabolism in intestinal mucosal cells play an important role in maintaining gut integrity and function, regulating endogenous synthesis of amino acids (citrulline, arginine, proline and alanine), and modulating the availability of dietary amino acids to extra-intestinal tissues (Wu, 2009). Because elevated levels of glutamate and aspartate in the circulation exert a neurotoxic effect, their extensive catabolism by the small intestine has important physiological significance. Our calculations further revealed that the ratios of most amino acids in diets relative to lysine differ markedly from those entering the portal vein from the small-intestinal lumen or appearing in plasma and body proteins (Table 5). The discrepancies in the patterns of amino acids between diets and body proteins are particularly pronounced for arginine, histidine, methionine, proline, glutamine, glycine, and serine. Therefore, ratios of these amino acids to lysine in body proteins may not be a reliable basis for estimating their optimal dietary requirements by growing pigs.

3. Regulatory role for amino acids in gene expression

Definition

Gene expression is defined as the translation of information encoded in a gene (deoxyribonucleic acid, DNA) into ribonucleic acid (RNA; including messenger, transfer, and ribosomal RNAs) and protein. This is a highly specific process in which a gene can be switched on or off in response to regulatory factors. Amino acids are not only substrates for protein synthesis but also affect one or more of the following steps: modification of chromatin (the complex of DNA and covering proteins, such as histones), transcription (synthesis of mRNA from DNA), post-transcriptional modification, RNA transport, mRNA degradation, translation (synthesis of protein/polypeptides from mRNA), and post-translational modifications (Brasse-Lagnel *et al.*, 2009; Bruhat *et al.*, 2009).

DNA transcription

Two of the most studied amino acids regarding regulation of gene expression in cells and animals are glutamine and arginine. Following initial studies of arginine regulation of argininosuccinate synthase expression in human cell lines (Jackson *et al.*, 1986), much progress has been made toward understanding the control of gene transcription by amino acids. It is now known that regulatory effects of certain amino acids on gene expression may be mediated by transcription factors (including the basic region/leucine zipper factors, activating transcription factors, and CCAAT/enhancer-binding protein), specific regulatory sequences (including amino acid response elements, nutrient sensing response elements, and multiple sites) in the promoter, and various *cis* elements distinct from amino acid response elements or nutrient sensing response elements (Brasse-Lagnel *et al.*, 2009).

Results of our microarray studies involving early-weaned pigs supplemented with or without glutamine indicated that early weaning resulted in increased (52-346%) expression of genes related to oxidative stress and immune activation, but decreased (35-77%) expression of genes related to macronutrient metabolism and proliferation of cells in the gut (Wang *et al.*, 2008a). Dietary glutamine supplementation increased intestinal expression (120-124%) of genes that are necessary for cell growth and removal of oxidants, while reducing (34-75%) expression of genes that promote oxidative stress and immune activation (Wang *et al.*, 2008a). These findings reveal coordinate alterations of gene expression in response to weaning and aid in providing molecular mechanisms for the beneficial effects of dietary glutamine supplementation to improve the nutritional status in young mammals.

Our work with obese rats supplemented with or without arginine revealed that high-fat feeding decreased mRNA levels for lipogenic enzymes, AMP-activated protein kinase (AMPK), glucose transporters, heme oxygenase 3, glutathione synthetase, superoxide dismutase 3, peroxiredoxin 5, glutathione peroxidase 3, and stress-induced protein, while increasing expression of carboxypeptidase-A, peroxisome proliferator activated receptor (PPAR)-α, caspase 2, caveolin 3, and diacylglycerol kinase (Jobgen *et al.*, 2009b). In contrast, dietary arginine supplementation reduced mRNA levels for fatty acid binding protein 1, glycogenin, protein phosphates 1B, caspases 1 and 2, and hepatic lipase, but increased expression of PPARγ, heme oxygenase 3, glutathione synthetase, insulin-like growth factor II, sphingosine-1-phosphate receptor, and stress-induced protein. Biochemical analysis revealed that oxidative stress in white adipose tissue of rats fed a high-fat diet was prevented by arginine supplementation (Jobgen *et al.*, 2009b). Collectively, these results indicate that supplementing high-fat diets with arginine results in differential regulation of gene expression to affect nutrient metabolism, redox state, fat accretion, and adipocyte differentiation in adipose tissue. Also, upregulation of mitochondrial biogenesis by arginine may provide another mechanism for enhanced oxidation of long-chain fatty acids and glucose in fat depots and other insulin-sensitive tissues (Jobgen *et al.*, 2006).

Histone modifications by methylation, acetylation and phosphorylation, as well as DNA methylation (occurring in the 5'-positions of cytosine residues within CpG dinucleotides throughout the mammalian genome) play an important role in epigenetic regulation of gene expression and physiological functions (Kouzarides, 2007). Amino acids (e.g., methionine, histidine, serine and glycine) are major donors of methyl groups that may affect activities of histone acetyltransferase (e.g., CREB-binding protein) and acetylase, as well as specific DNA methyltransferases (Oommen *et al.*, 2005). Upon acetylation of histones, DNA is dissociated from histones so that transcription will proceed. In contrast, DNA methylation and histone deacetylation result in highly dense packing of DNA and, therefore, silence gene expression. RNA polymerase catalyzes the transcription of a gene to mRNA. This process can be regulated by one or more of the following mechanisms: 1) alteration of specificity of RNA polymerase for promoters; 2) binding of repressors to non-coding DNA sequences that are near or overlap the promoter region; and 3) transcription factors (e.g., upregulation, downregulation, coactivators, and corepressors) (Waterland and Jirtle, 2004). Post-transcriptional regulation is mediated by capping (changing the 5' end of the mRNA to a 3' end, conferring protection of the mRNA from 5' exonuclease); splicing (removing introns and joining of exons); addition of poly(A) tail (poly-adenylation) to confer protection of mRNA from 3' exonuclease. Covalent modifications of DNA and core histones provide a basis for epigenetics, which is defined as stable alterations in gene expression without changes in the underlying DNA sequence (Kouzarides, 2007). Epigenetic changes may remain through cell divisions and, therefore, may be carried forward to subsequent generations (Waterland and Jirtle, 2004). This notion of transgenerational effects of nutrients has very important implications for animal production.

mRNA translation

Initiation of mRNA translation is a key event in the regulation of protein synthesis. The mammalian target of rapamycin (mTOR), a highly conserved serine/threonine protein kinase, is the master regulator of translation (Davis *et al.*, 2009); mTOR is also known as FK506 binding protein 12-rapamycin associated protein 1 (FRAP1). The mTOR system consists of 1) mTOR complex-1 (mTOR, raptor, and G protein β-subunit-like protein) and 2) mTOR complex-2 (mTOR, rictor, mitogen-activated-protein-kinase-associated protein 1, and G protein β-subunit-like protein) (Shaw, 2008). Amino acids (e.g., glutamine, arginine, and leucine) stimulate the phosphorylation of mTOR in a cell-specific manner, leading to phosphorylation of S6K and 4EBP1 proteins and, therefore, the formation of translation initiation complex (Yao *et al.*, 2008; Rhoads and Wu, 2009). Besides mTOR signaling, arginine can regulate gene expression and activity of AMPK, another important protein in nutrient sensing that modulates oxidation of energy substrates and insulin sensitivity (Jobgen *et al.*, 2006). Additionally, glutamine is known to activate several signaling pathways, including extracellular signal-related kinase, Jun kinase, mitogen-activated protein kinase, protein kinase A, nuclear receptors, zinc fingers proteins, and helix-turn-helix proteins (Brasse-Lagnel *et al.*, 2009).

Results from cell culture studies have led to three models to explain how amino acids can regulate protein synthesis in animal cells (Figure 2). It should be borne in mind that these *in vitro* experiments were conducted under conditions of either complete absence of an amino acid or its presence at a high concentration (e.g., 2 mM; at least 10 times plasma concentrations for most amino acids) in culture medium (Wang *et al.*, 2008b). Therefore, it is unknown whether the findings are physiologically relevant to animals. Nonetheless, some of these hypotheses are supported by results from *in vivo* studies (e.g., Wang *et al.*, 2008a; Jobgen *et al.*, 2009a,b; Rhoads and Wu, 2009). Moreover, the potential for certain amino acids (both essential and nonessential) to activate gene transcription and regulate mRNA translation is truly

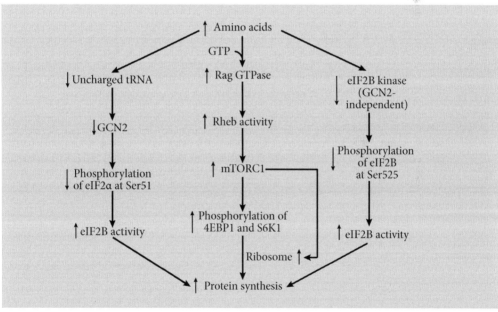

Figure 2. Models for regulatory roles of amino acids in protein synthesis. Provision of sufficient amounts of amino acids promotes the binding of GTP to Rag GTPase leading to mTORC1 activation, while reducing concentrations of uncharged tRNA and activity of the GCN2-independent eIF2B kinase leading to increased activity of eIF2B. All of these changes stimulate the initiation of protein synthesis in cells. Abbreviations: eIF2α, eukaryotic translation initiation factor 2B; eIF2B, eukaryotic translation initiation factor 2B; GCN2, general control nonderepressible protein 2; mTORC1, mammalian target of rapamycin complex 1; S6K1, ribosomal protein S6 kinase-1; 4EBP1, eIF4E-binding protein-1. ↑ increase in concentration or activity; ↓ decrease in concentration or activity.

remarkable (Shaw, 2008). This argues that the traditional definition of essential or nonessential amino acids solely on the basis of needs for these nutrients as building blocks of proteins has severe conceptual limitations.

Synthesis of NO, CO and H₂S

Nitric oxide (NO), carbon monoxide (CO), and hydrogen sulfide (H$_2$S) are three physiological gases that have received much attention in recent years (Li *et al.*, 2009). They are synthesized from amino acids. NO is synthesized from arginine by one of the three isoforms of NO synthase (NOS): neuronal NOS (nNOS; also known as NOS1), inducible NOS (iNOS; also known as NOS2), and endothelial NOS (eNOS; also known as NOS3) (Wu and Morris, 1998). nNOS, eNOS, and iNOS are localized primarily in cytoplasm and mitochondria, plasma membrane and cytoplasm, and cytoplasm, respectively. nNOS and eNOS are constitutively expressed in a cell-specific manner and produce low levels of NO, whereas iNOS is induced by certain immunological stimuli (including LPS and inflammatory cytokines) and generates a large amount of NO (Jobgen *et al.*, 2006). However, all isoforms of NOS depend on NADPH, calmodulin, FAD, FMN and tetrahydrobiopterin as essential cofactors for their enzymatic activities. In addition, nNOS and eNOS, but not iNOS, require calcium for generation of NO.

CO is produced from the degradation of heme by membrane-bound heme oxygenase (HO or HMOX) in the body (Durante *et al.*, 2006). The HO system depends on NADPH-cytochrome 450 for its catalytic activity. In cells, heme is synthesized from glycine and succinyl CoA via a series of reactions involving both cytoplasm and mitochondria, and the carbon of CO is derived from the methyl carbon (carbon-2) of glycine. There are three isoforms of HO: HO1 (also known as heat shock protein 32), HO2, and HO3. HO1 is a highly inducible protein, and its activity can increase up to 100-fold when cells are treated with oxidants, endotoxins, or inflammatory cytokines. In contrast, HO2 is constitutively expressed and its activity is not altered by these agents. H$_2$S is produced from cysteine (a metabolite of methionine) via multiple pathways, including cystathionine β-synthase (CBS), cystathionine γ-lyase (also known as cystathionase; CSE), and β-mercaptopyruvate sulfurtransferase. CBS and CSE are pyridoxal phosphate-dependent cytosolic enzymes responsible for production of most H$_2$S in animals (Stipanuk *et al.*, 2006). CBS is Ca^{2+} and calmodulin-dependent, whereas CSE is inducible by endotoxin and likely inflammatory cytokines (Kamoun, 2004).

Many amino acids have been reported to regulate the production of NO, CO, and H$_2$S in a cell-dependent manner (Wu and Meininger, 2002; Li *et al.*, 2009). For example, arginine, citrulline, glutamate, glycine, taurine, and γ-aminobutyrate increase NO synthesis by constitutive NO synthase in endothelial cells or brain, whereas low protein intake, glucosamine, glutamine, and lysine inhibit NO generation by both constitutive and inducible NO synthases. In addition, high protein intake, arginine, glutamine, glutamate, alanine, taurine, methionine, and glycine all promote CO synthesis by heme oxygenase in endothelial cells and nonvascular tissues, but N-acetyl-cysteine attenuates CO formation in injured brain and vascular smooth muscle cells. Furthermore, H_2S production is stimulated by high protein intake, arginine, cysteine, methionine, glycine, S-adenosylmethionine, N-acetyl-glutamate, and glutamate (0.1 mM), but is inhibited by aspartate and higher levels of glutamate (1-3 mM), in various cell types. Either imbalance or antagonism among amino acids affects the generation of one or more of these gaseous molecules and, therefore, protein nutrition, in animals (Li *et al.*, 2009). This should be taken into consideration in formulating swine diets.

Physiological effects of NO, CO and H₂S

NO is a highly reactive free radical. In contrast, CO, H_2S and SO_2 are strong reducing agents. However, they are all water-soluble colorless molecules that easily penetrate biological membranes. Therefore, these gases exert their effects on cells independent of membrane receptors. Because of their chemical properties, effects of NO, CO, H_2S and SO_2 depend on their respective concentrations in cells. At pathological levels, they are extremely destructive to all cell types. At physiological levels, NO, CO and SO_2 activate guanylyl cyclase to generate cGMP which elicits a variety of responses via cGMP-dependent protein kinases that include relaxation of vascular smooth muscle cells, hemodynamics, neurotransmission, and cell metabolism) (Li *et al.*, 2007; 2009). The actions of the gases may also involve cGMP-independent mechanisms (e.g., inhibition of ornithine decarboxylase, protein modification, and redox state). Emerging evidence shows that H_2S is a crucial regulator of both neurological function and endothelium-dependent relaxation through cGMP-independent mechanisms involving stimulation of membrane K_{ATP} channels and intracellular cAMP signaling (Kimura *et al.*, 2005). Additionally, physiological levels of NO, CO and H_2S confer cytoprotective and immunomodulatory effects.

4. Applications of functional amino acids to swine nutrition and production

Dietary L-glutamine supplementation improves survival and growth of young pigs

Piglets with intrauterine growth retardation (IUGR) suffer from necrotizing enterocolitis, a major cause of death in neonates (Wu *et al.*, 2006). They are also more susceptible to infections and have a high rate of mortality. Based on its versatile roles in intestinal physiology and its safety, L-glutamine (1 g/kg body weight per day) has been administered orally to IUGR piglets to effectively improve their survival and growth, particularly in response to endotoxin treatment (Haynes *et al.*, 2009). Intestinal atrophy in weanling piglets is another significant problem in swine nutrition and production. This digestive abnormality results from multiple factors, including immunological challenges, oxidative stress, apoptosis, and insufficient energy provision. We found that dietary supplementation with 1% L-glutamine prevented jejunal atrophy during the first week postweaning and increased the gain:feed ratio by 25% during the second week postweaning (Wu *et al.*, 1996b; Wang *et al.*, 2008a). Post-weaning pigs fed a milk-based or a corn- and soybean meal-based diet tolerated up to 1.12% supplemental L-glutamine (calculated on a dry matter basis in the diet) for at least 3 months without any adverse effect or toxicity. In all of our experiments, dietary supplementation with up to 1.12% L-glutamine (dry matter basis) caused no sickness or death in any pig. No side effects of glutamine supplementation (up to 1.12% in the diet on dry matter basis) were observed in pigs within at least 3 months after termination of a two-week, one-month or three-month period of supplementation. Our findings led to the commercial development and availability of feed-grade glutamine (AminoGut) by Ajinomoto Co., Inc. for use in swine diets (www.ajinomoto.com).

Dietary L-proline supplementation promotes growth of young pigs

Proline metabolism in pigs differs markedly with developmental stage (Wu, 1997; Bertolo and Burrin, 2008). Proline is a nutritionally essential amino acid for young mammals (Ball *et al.*, 1986), because of inadequate endogenous synthesis via the arginase and P5C synthase pathways relative to needs (Wu and Morris, 1998). Thus, supplementing 0, 0.35, 0.7, 1.05, 1.4, and 2.1% L-proline to a proline-free chemically defined diet containing 0.48% L-arginine and 2% L-glutamate dose-dependently improved daily weight gains (from 342 to 411 g per day) and feed efficiency (gram feed/gram gain; from 1.66 to 1.35) of young pigs, while reducing concentrations of urea in plasma by one-half (Kirchgessner *et al.*, 1995). Notably, increasing the dietary content of L-proline from 0.0

and 2.1% enhanced daily nitrogen retention from 1.27 to 1.53 g/kg body weight$^{0.75}$ (metabolic weight).

Dietary L-arginine supplementation enhances growth in milk-fed piglets

Data from artificial rearing systems indicate that the biological potential for piglet growth is at least 400 g/day (average from birth to 21 days of age) or $\geq$ 74% greater than that for sow-reared piglets (230 g/d) and that suckling piglets start to exhibit submaximal growth from Day 8 after birth (Boyd et al., 1995). We discovered that arginine deficiency is a major factor limiting maximal growth of milk-fed piglets (Wu et al., 2004). Dietary supplementation with 0.2% and 0.4% L-arginine to 7- to 21-day-old milk-fed pigs (artificially reared on a liquid-milk feeding system) dose-dependently enhanced plasma arginine concentrations (30% and 61%), reduces plasma ammonia levels (20% and 35%), and increased weight gain (28% and 66%) (Kim and Wu, 2004). Mateo et al. (2008) recently reported that supplementing 1.0% arginine-HCl to the diet for lactating sows increased milk production and piglet growth possibly due to increased angiogenesis in and blood flow to the mammary gland. Provision of L-arginine or arginine-rich rice protein concentrate to either sow-reared or weanling pigs is highly effective in improving their growth performance and immune function (Yao et al., 2008; Hou et al., 2008; Liu et al., 2009). Such commercial products are now available to pork producers worldwide.

Dietary L-arginine supplementation enhances spermatogenesis in boars

Concentrations of polyamines (products of arginine catabolism) are relatively high in porcine seminal fluid (90 µM) with the ratio of putrescine:spermidine:spermine being 1:5.75:5.78 (Wu et al., 2009). Dietary supplementation with 1% L-arginine-HCl to sexually active boars for 30 days did not affect the volume of ejaculated semen but enhanced concentrations of arginine, proline, ornithine, and polyamines in seminal fluid by 43%, 41%, 56%, and 63%, respectively, compared with the control group. Importantly, arginine supplementation increased sperm counts by 18% and sperm motility by 7.6% compared with the control group (Wu et al., 2009). The underlying mechanisms may involve augmented synthesis of polyamines and NO that are essential to spermatogenesis and sperm viability. Dietary supplementation with arginine may provide a novel means to improve fertility in boars and other male breeding animals, particularly under stressful conditions.

Dietary L-arginine supplementation improves pregnancy outcome in pigs

Among livestock species, pigs suffer the greatest prenatal loss (up to 50%) due to a suboptimal intrauterine environment, which may include inadequate uterine secretions and sub-optimal nutrition (Wu *et al.*, 2006). This problem is even more severe in modern highly prolific pigs than in breeds used in the swine industry 25 or 30 years ago due to selection for high ovulation rate (Pond *et al.*, 1981; Vonnahme *et al.*, 2002). Thus, prolific gilts or sows ovulate 20 to 30 oocytes, but deliver only 9 to 15 piglets at term (Town *et al.*, 2005). Current restricted feeding programs for gestating swine to prevent excessive maternal weight gains result in inadequate provision of arginine from mother to fetuses (Wu *et al.*, 1999). Unfortunately, the current version of NRC (1998) recommends little or no requirement for dietary arginine by gestating gilts or sows, because previous studies with low prolific swine indicated that dietary arginine was not needed to maintain a positive nitrogen balance (Easter and Baker, 1976) and that a lack of arginine in the diet did not affect litter size or fetal growth (Easter and Baker, 1974).

We discovered an unusually high abundance of arginine in porcine allantoic fluid during early gestation (Wu *et al.*, 1996a). In fact, arginine plus ornithine account for 50% and 55% of the total α-AA nitrogen in porcine allantoic fluid on days 40 and 45 of gestation, respectively. Therefore, we hypothesized that arginine plays an important role in conceptus (embryo/fetus and associated extra-embryonic membranes) survival and growth and that increasing dietary provision of arginine beyond that from a typical corn- and soybean meal-based diet may be an effective means to enhance circulating levels and improve pregnancy outcome in pigs. Several lines of experimental evidence support this hypothesis. First, dietary supplementation with 1.0% arginine-HCl between days 30 and 114 of gestation increased the number of live-born piglets by two and litter birthweight by 24% (Mateo *et al.*, 2007). Second, dietary supplementation with 1% L-arginine to gilts or sows between days 14 and 28 of gestation increased the number of live-born piglets by approximately one at birth (Campbell, 2009; Ramaekers *et al.*, 2006). Third, arginine supplementation (1% in the diet) between days 14 and 28 of gestation increased the number of fetuses on day 70 by three per litter in hyper-ovulating gilts (Berard *et al.*, 2009).

Dietary L-arginine supplementation improves growth performance and meat quality in growing-finishing pigs

Rates of muscle glycolysis affect the extent of post-mortem pH decline and its water-holding capacity. A rapid pH decline in muscle during the first 60 min post-mortem can lead to the pale-soft-exudative pork condition (Bendall and Swatland, 1988). Results of our study demonstrated that supplementing the diet for growing-finishing pigs with 1% L-arginine increased glycogen content by 42% and reduced lactate content by 37% in longissimus muscle (due to reduced rates of glycogenolysis and glycolysis, respectively), while enhancing muscle pH at 45 min post-mortem by 0.32 (Tan *et al.*, 2009). Ma and coworkers (2008) also reported that dietary L-arginine supplementation enhanced tissue anti-oxidative capacity in growing-finishing pigs and pork quality. These novel and very important findings indicate that arginine supplementation may be a new effective means to ameliorate the deleterious effects of rapid pH decline in post-mortem pork muscle during early postmortem. Additionally, we found that L-arginine supplementation increased body weight gain by 6.5% and carcass skeletal-muscle content by 5.5%, while decreasing carcass fat content by 11%. These results support the idea that dietary arginine supplementation beneficially promotes muscle gain and reduces body fat accretion in growing-finishing pigs.

5. Conclusion

The traditional classification of amino acids as nutritionally essential or nonessential has major conceptual limitations. It is also unfortunate that the current version of NRC does not recommend dietary requirements of 'nonessential amino acids' by neonatal, postweaning, growing-finishing, or gestating pigs. However, emerging evidence shows that these amino acids, particularly glutamine and arginine, play important roles in regulating gene expression at both transcriptional and translational levels in animals. Additionally, there is growing recognition that amino acids (including nonessential amino acids) participate in cell signaling via mTOR, AMPK, extracellular signal-related kinase, Jun kinase, mitogen-activated protein kinase, and gases (NO, CO and H_2S). Exquisite integration of these regulatory networks has profound effects on cell proliferation, differentiation, metabolism, homeostasis, survival, and function. Unquestionably, recent advances in understanding of functional amino acids are transforming the practice of swine nutrition. Availability of feed-grade amino acids (e.g., arginine, glutamine, leucine, and proline) is expected to improve the efficiency and quality of pork production worldwide.

Acknowledgments

We thank all the personnel in our laboratories for technical support, Kenton Lilie for assistance with animal husbandry, as well as Merrick Gearing and Frances Mutscher for manuscript preparation. This work was supported by the National Research Initiative Competitive Grant

Animal Reproduction Program (No. 2006-35203-17199 and 2008-35203-19120) and Animal Growth and Nutrient Utilization Program (2008-35206-18762, 2008-35206-18764 and 2009-35206-05211) from the USDA Cooperative State Research, Education, and Extension Service, Texas AgriLife Research (H-82000), and North Carolina Agricultural Research Service (404050).

References

Baker, D.H., 2005. Comparative nutrition and metabolism: explication of open questions with emphasis on protein and amino acids. Proceedings of the National Academy of Science USA 102: 17897-17902.

Ball, R.O., Atkinson, J.L., Bayley, H.S., 1986. Proline as an essential amino acid for the young pig. British Journal of Nutrition 55: 659-668.

Bendall, J.K., Swatland, H.J., 1988. A review of the relationship of pH with physical aspects of pork quality. Meat Science 24: 85-96.

Berard, J., Kreuzer, M., Bee, G., 2009. Effect of dietary arginine supplementation to sows on litter size, fetal weight and myogenesis at d 75 of gestation. Journal of Animal Science 87 (E-Suppl. 3): 30 (Abstract).

Bergen, W.G., Wu, G., 2009. Intestinal nitrogen recycling and utilization in health and disease. Journal of Nutrition 139: 821-825.

Boyd, R.D., Kensinger, R.S., Harrell, R.J., Bauman, D.E., 1995. Nutrient uptake and endocrine regulation of milk synthesis by mammary tissues of lactating sows. Journal of Animal Science 73 (Suppl. 2): 36-56.

Brasse-Lagnel, C., Lavoinne, A., Husson, A., 2009. Control of mammalian gene expression by amino acids, especially glutamine. FEBS Journal 276: 1826-1844.

Bruhat, A., Cherasse, Y., Chaveroux, C., Maurin, A.C., Jousse, C., Fafournous, P., 2009. Amino acids as regulators of gene expression in mammals: molecular mechanisms. BioFactors 35: 249-257.

Bertolo, R.F., Burrin, D.G., 2008. Comparative aspects of tissue glutamine and proline metabolism. Journal of Nutrition 138: 2032S-2039S.

Campbell, R., 2009. Pork CRC – NZ Seminar Series: Arginine and reproduction. Available at: http://www/nzpib.co.nz. Accessed July 28, 2009.

Chen, L., Yin, Y., Jobgen, W.J., Jobgen, S.C., Knabe, D.A., Hu, W., Wu, G., 2007. In vitro oxidation of essential amino acids by intestinal mucosal cells of growing pigs. Livestock Science 109: 19-23.

Chen, L., Li, P., Wang, J., Li, X., Gao, H., Yin, Y., Hou, Y., Wu, G., 2009. Catabolism of essential amino acids in developing porcine enterocytes. Amino Acids 37: 143-152.

Davis, T.A., Fiorotto, M.L., 2009. Regulation of muscle growth in neonates. Current Opinion in Nutrition and Metabolic Care 12: 78-85.

Durante, W., Johnson, F.K., Johnson, R.A., 2006. Role of carbon monoxide in cardiovascular function. Journal of Cellular and Molecular Medicine 10: 672-686.

Easter, R.A., Baker, D.H., 1974. Arginine: a dispensable amino acid for postpuberal growth and pregnancy of swine. Journal of Animal Science 39: 1123-1128.

Easter, R.A., Baker, D.H., 1976. Nitrogen metabolism and reproductive response of gravid swine fed an arginine-free diet during the last 84 days of gestation. Journal of Nutrition 106: 636-641.

Flynn, N.E., Wu, G., 1996. An important role for endogenous synthesis of arginine in maintaining arginine homeostasis in neonatal pigs. American Journal of Physiology 271: R1149-R1155.

Fuller, M.F., Reeds, P.J., 1998. Nitrogen cycling in the gut. Annual Review of Nutrition 18: 385-411.

Haynes, T.E., Li, P., Li, X.L., Shimotori, K., Sato, H., Flynn, N.E., Wang, J.J., Knabe, D.A., Wu, G., 2009. L-Glutamine or L-alanyl-L-glutamine prevents oxidant- or endotoxin-induced death of neonatal enterocytes. Amino Acids 37: 131-142.

Hou, Z.P., Yin, Y.L., Huang, R.L., Li, T.J., Hou, Y.Q., Liu, Y.L., Wu, X., Liu, Z.Q., Wang, W., Xiong, H., Wu, G.Y., Tan, L.X., 2008. Rice protein concentrate partially replaces dried whey in the diet for early-weaned piglets and improves their growth performance. Journal of the Science of Food and Agriculture 88: 1187-1193.

Jackson, M.J., O'Brien, W.E., Beaudet, A.L., 1986. Arginine-mediated regulation of an argininosuccinate synthetase minigene in normal and canavanine-resistant human cells. Molecular and Cellular Biology 6: 2257-2261.

Jobgen, W.S., Fried, S.K., Fu, W.J., Meininger, C.J., Wu, G., 2006. Regulatory role for the arginine-nitric oxide pathway in metabolism of energy substrates. Journal of Nutritional Biochemistry 17: 571-588.

Jobgen, W.J., Meininger, C.J., Jobgen, S.C., Li, P., Lee, M.-J., Smith, S.B., Spencer, T.E., Fried, S.K., Wu, G., 2009a. Dietary L-arginine supplementation reduces white-fat gain and enhances skeletal muscle and brown fat masses in diet-induced obese rats. Journal of Nutrition 139: 230-237.

Jobgen, W., Fu, W.J., Gao, H., Li, P., Meininger, C.J., Smith, S.B., Spencer, T.E.,, Wu, G., 2009b. High fat feeding and dietary L-arginine supplementation differentially regulate gene expression in rat white adipose tissue. Amino Acids 37: 187-198.

Kamoun, P., 2004. Endogenous production of hydrogen sulfide in mammals. Amino Acids 26: 243-254.

Kim, S.W., Wu, G., 2004. Dietary arginine supplementation enhances the growth of milk-fed young pigs. Journal of Nutrition 134: 625-630.

Kim, S.W., Wu, G., Baker, D.H., 2005. Ideal protein and amino acid requirements by gestating and lactating sows. Pig News and Information 26: 89N-99N.

Kim, S.W., Hurley, W.L., Wu, G., Ji, F., 2009. Ideal amino acid balance for sows during gestation and lactation. Journal of Animal Science 87: E123-132.

Kimura, H., Nagai, Y., Umemura, K., Kimura, Y., 2005. Physiological roles of hydrogen sulfide: synaptic modulation, neuroprotection, and smooth muscle relaxation. Antioxidants and Redox Signaling 7: 795-803.

Kirchgessner, M., Fickler, J., Roth, F.X., 1995. Effect of dietary proline supply on N-balance of piglets. 3. Communication on the importance of nonessential amino acids for protein retention. Journal of Animal Physiology and Animal Nutrition 73: 57-65.

Kong, X.F., Yin, Y.L., He, Q.H., Yin, F.G., Liu, H.J., Li, T.J., Huang, R.L., Geng, M.M., Ruan, Z., Deng, Z.Y., Xie, M.Y., Wu, G., 2008. Dietary supplementation with Chinese herbal powder enhances ileal digestibilities and serum concentrations of amino acids in young pigs. Amino Acids doi: 10.1007/s00726-008-0176-9.

Kouzarides, T., 2007. Chromatin modifications and their function. Cell 128: 693-705.

Li, P., Yin, Y.L., Li, D.F., Kim, S.W., Wu, G., 2007. Amino acids and immune function. British Journal of Nutrition 98: 237-252.

Li, X., Bazer, F.W., Gao, H., Jobgen, W., Johnson, G.A., Li, P., McKnight, J.R., Satterfield, M.C., Spencer, T.E., Wu, G., 2009. Amino acids and gaseous signaling. Amino Acids 37: 65-78.

Liu, Y., Huang, J., Hou, Y., Zhu, H., Zhao, S., Ding, B., Yin, Y., Yi, G., Shi, J., Fan, W. 2008. Dietary arginine supplementation alleviates intestinal mucosal disruption induced by Escherichia coli lipopolysaccharide in weaned pigs. British Journal of Nutrition 100: 552-560.

Ma, X., Lin, Y., Jiang, Z., Zheng, C., Zhou, G., Yu, D., Cao, T., Wang, J., Chen, F., 2008. Dietary arginine supplementation enhances antioxidative capacity and improves meat quality of finishing pigs. Amino Acids doi:10.1007/s00726-008-0213-8.

Mateo, R.D., Wu, G., Bazer, F.W., Park, J.C., Shinzato, I., Kim, S.W., 2007. Dietary L-arginine supplementation enhances the reproductive performance of gilts. Journal of Nutrition 137: 652-656.

Mateo, R.D., Wu, G., Moon, H.K., Carroll, J.A., Kim, S.W., 2008. Effects of dietary arginine supplementation during gestation and lactation on the performance of lactating primiparous sows and nursing piglets. Journal of Animal Science 86: 827-835.

Mavromichalis, I., Parr, T.M., Gabert, V.M., Baker, D.H., 2001. True ileal digestibility of amino acids in sow's milk for 17-day-old pigs. Journal of Animal Science 79: 707-713.

Murphy, J.M., Murphy, S.J., Ball, R.O., 1996. Proline is synthesized from glutamate during intragastric infusion but not during intravenous infusion in neonatal piglets. Journal of Nutrition 126: 878-886.

National Research Council (NRC), 1998. Nutrient Requirements of Swine (10th edition). National Academy Press, Washington DC, USA, 189 pp.

Oommen, A.M., Griffin, J.B., Sarath, G., Zempleni, J., 2005. Roles for nutrients in epigenetic events. Journal of Nutritional Biochemistry 16: 74-77.

Pond, W.G., Yen, J.T., Mauer, R.R., Christenson, R.K., 1981. Effect of doubling daily energy intake during the last two weeks of pregnancy on pig birth weight, survival and weaning weight. Journal of Animal Science 52: 535-541.

Ramaekers, P., Kemp, B., Van der Lende, T., 2006. Progenos in sows increases number of piglets born. Journal of Animal Science 84 (Suppl. 1): 394 (Abstract).

Reeds, P.J., Burrin, D.G., 2001. Glutamine and the bowel. Journal of Nutrition 131: 2505S-2508S.

Reeds, P.J., Burrin, D.G., Jahoor, F., Wykes, L., Henry, J., Frazer, M.E., 1996. Enteral glutamate is almost completely metabolized in first pass by the gastrointestinal tract of infant pigs. American Journal of Physiology 270: E413-E418.

Reeds, P.J., Burrin, D.G., Stoll, B., Jahoor, F., Wykes, L., Henry, J., Frazer, M.E., 1997. Enteral glutamate is the preferential source for mucosal glutathione synthesis in pigs. American Journal of Physiology 273: E408-E415.

Rhoads, J.M., Wu, G., 2009. Glutamine, arginine, and leucine signaling in the intestine. Amino Acids 37: 111-122.

Self, J.T., Spencer, T.E., Johnson, G.A., Hu, J., Bazer, F.W., Wu, G., 2004. Glutamine synthesis in the developing porcine placenta. Biology of Reproduction 70: 1444-1451.

Shaw, R.J., 2008. mTOR signaling: RAG GTPases transmit the amino acid signal. Trends in Biochemical Sciences 33: 565-568.

Stipanuk, M.H., Dominy, J.E., Lee, J.I., Coloso, R.M., 2006. Mammalian cysteine metabolism: new insights into regulation of cysteine metabolism. Journal of Nutrition 136: 1652S-1659S.

Stoll, B., Burrin, D.G., Henry, J., Yu, H., Jahoor, F., Reeds, P.J., 1999. Substrate oxidation by the portal drained viscera of fed piglets. American Journal of Physiology 277: E168-E175.

Stoll, B., Burrin, D.G., 2006. Measuring splanchnic amino acid metabolism in vivo using stable isotopic tracers. Journal of Animal Science 84: E60-E72.

Suryawan, A., O'Connor, P.M.J., Bush, J.A., Nguyen, H.V., Davis, T.A., 2009. Differential regulation of protein synthesis by amino acids and insulin in peripheral and visceral tissues of neonatal pigs. Amino Acids 37: 105-110.

Tan, B., Yin, Y.L., Liu, Z.Q., Li, X.G., Xu, H.J., Kong, X.F., Huang, R.L., Tang, W.J., Shinzato, I., Smith, S.B., Wu, G., 2009. Dietary L-arginine supplementation increases muscle gain and reduces body fat mass in growing-finishing pigs. Amino Acids 37: 169-175.

Town, S.C., Patterson, J.L., Pereira, C.Z., Gourley, G., Foxcroft, G.R., 2005. Embryonic and fetal development in a commercial dam-line genotype. Animal Reproduction Science 85: 301-316.

Vonnahme, K., Wilson, M.E., Foxcroft, G.R., Ford, S.P., 2002. Impacts on conceptus survival in a commercial swine herd. Journal of Animal Science 80: 553-559.

Wang, J.J., Chen, L.X., Li, P., Li, X.L., Zhou, H.J., Wang, F.L., Li, D.F., Yin, Y.L., Wu, G., 2008a. Gene expression is altered in piglet small intestine by weaning and dietary glutamine supplementation. Journal of Nutrition 138: 1025-1032.

Wang, X.M., Fonseca, B.D., Tang, H., Liu, R., Elia, A., Clemens, M.J., Bommer, U.A., Proud, C.G., 2008b. Re-evaluating the role of proposed modulators of mammalian target of rapamycin complex 1 (mTORC1) signaling. Journal of Biological Chemistry 283: 30482-30492.

Waterland, R.A., Jirtle, R.L., 2004. Early nutrition, epigenetic changes at transposons and imprinted genes, and enhanced susceptibility to adult chronic diseases. Nutrition 20: 63-68.

Wu, G., 1997. Synthesis of citrulline and arginine from proline in enterocytes of postnatal pigs. American Journal of Physiology 272: G1382-G1390.

Wu, G., 1998. Intestinal mucosal amino acid catabolism. Journal of Nutrition 128: 1249-1252.

Wu, G., 2009. Amino acids: metabolism, functions, and nutrition. Amino Acids 37: 1-17.

Wu, G., Knabe, D.A., 1994. Free and protein-bound amino acids in sow's colostrum and milk. Journal of Nutrition 124: 415-424.

Wu, G., Borbolla, A.G., Knabe, D.A., 1994. The uptake of glutamine and release of arginine, citrulline and proline by the small intestine of developing pigs. Journal of Nutrition 124: 2437-2444.

Wu, G., Knabe, D.A., 1995. Arginine synthesis in enterocytes of neonatal pigs. American Journal of Physiology 269: R621-R629.

Wu, G., Bazer, F.W., Tuo, W., Flynn, S.P., 1996a. Unusual abundance of arginine and ornithine in porcine allantoic fluid. Biology of Reproduction 54: 1261-1265.

Wu, G., Meier, S.A., Knabe, D.A., 1996b. Dietary glutamine supplementation prevents jejunal atrophy in weaned pigs. Journal of Nutrition 126: 2578-2584.

Wu, G., Knabe, D.A., Flynn, N.E., Yan, W., Flynn, S.P., 1996c. Arginine degradation in developing porcine enterocytes. American Journal of Physiology 271: G913-G919.

Wu, G., Morris, S.M., Jr., 1998. Arginine metabolism: nitric oxide and beyond. Biochemical Journal 336: 1-17.

Wu, G., Ott, T.L., Knabe, D.A., Bazer, F.W., 1999. Amino acid composition of the fetal pig. Journal of Nutrition 129: 1031-1038.

Wu, G., Flynn, N.E., Knabe, D.A., 2000. Enhanced intestinal synthesis of polyamines from proline in cortisol-treated piglets. American Journal of Physiology 279: E395-E402.

Wu, G., Meininger, C.J., 2002. Regulation of nitric oxide synthesis by dietary factors. Annual Review of Nutrition 22: 61-86.

Wu, G., Knabe, D.A., Kim, S.W., 2004. Arginine nutrition in neonatal pigs. Journal of Nutrition 134: 2783S-2390S.

Wu, G., Bazer, F.W., Hu, J., Johnson, G.A., Spencer, T.E., 2005. Polyamine synthesis from proline in the developing porcine placenta. Biology of Reproduction 72: 842-850.

Wu, G., Bazer, F.W., Wallace, J.M., Spencer, T.E., 2006. Intrauterine growth retardation: Implications for the animal sciences. Journal of Animal Science 84: 2316-2337.

Wu, G., Bazer, F.W., Cudd, T.A., Jobgen, W.S., Kim, S.W., Lassala, A., Li, P., Matis, J.H., Meininger, C.J., Spencer, T.E., 2007. Pharmacokinetics and safety of arginine supplementation in animals. Journal of Nutrition 137: 1673S-1680S.

Wu, G., Bazer, F.W., Davis, T.A., Kim, S.W., Li, P., Rhoads, J.M., Satterfield, M.C., Smith, S.B., Spencer, T.E., Yin, Y.L., 2009. Arginine metabolism and nutrition in growth, health and disease. Amino Acids 37: 153-168.

Yao, K., Yin, Y.L., Chu, W.Y., Liu, Z.Q., Deng, D., Li, T.J., Huang, R.L., Zhang, J.S., Tan, B.E., Wang, W., Wu, G., 2008. Dietary arginine supplementation increases mTOR signaling activity in skeletal muscle of neonatal pigs. Journal of Nutrition 138: 867-872.

Yen, J.T., Killefer, J., 1987. A method for chronically quantifying net absorption of nutrients and gut metabolites into hepatic portal vein in conscious swine. Journal of Animal Science 64: 923-934.

Glucose metabolism in reproductive sows

Rosemarijn Gerritsen[1], Paul Bikker[1,2] and Piet van der Aar[1]
[1]*Schothorst Feed Research B.V., P.O. Box 533, 8200 AM, Lelystad, The Netherlands; rgerritsen@schothorst.nl*
[2]*Wageningen UR, Livestock Research, P.O. Box 65, 8200 AB Lelystad, The Netherlands*

Abstract

Glucose is an important energy source for sows throughout the production cycle. It directly affects oocyte maturation, glycogen deposition in the fetal piglet during gestation and it is a precursor of lactose and therefore important for milk production. In this paper the effects of glucose metabolism on performance of gestating and lactating sows are discussed. The developing fetus and the uterus rely on glucose as their main source of energy. Studies have shown that increasing the level of starch in diets of gestating sows can increase birth weight of piglets and possibly also affect vitality in new born piglets. At the end of gestation, the fetuses grow fast and the sow develops a reversible insulin resistance, in order to increase the glucose transport to the fetuses. Glucose transport can be increased because during this period of insulin resistance, lowering insulin sensitivity results in higher blood glucose levels for a longer period of time. Studies have shown that insulin resistance in sows can be affected by e.g. the dietary fatty acid profile. Possible effects of other dietary nutrients such as protein need further investigation. It has been shown that piglet survival, especially of the lighter piglets within a litter, can be improved by feeding medium chain triglycerides to gestating sows. Increasing insulin resistance, however, seems to have a negative effect on piglet survival. In contrast to other species that maintain insulin resistance during lactation in order to stimulate glucose transport to the udder, insulin resistance in sows only seems present during the end of the gestation period. However, glucose requirements of the lactating sow are high as it is the main precursor for lactose. About 140 grams of glucose is required to produce 1 liter of milk. Higher levels of starch in the lactation diet result in a higher milk production with a higher lactose content but lower fat and energy contents. Piglet gain, however is improved. These positive effects of starch are the result of a higher level of available glucose. The level of dietary glucose has been also found to affect sow reproduction. Increasing the level of glucose in the sow diet during the weaning to oestrus interval positively affects litter size and within-litter variation. At present is not clear how glucose affects reproduction, but the general thought is that it acts via insulin.

In conclusion, dietary glucose is of importance throughout the entire reproductive cycle of the sow. Therefore, requirements for glucose in the modern sow during each of the different stages of the reproductive cycle should be further studied.

1. Introduction

For optimal reproductive performance, the sow requires sufficient nutrients. The nutritional requirements for caloric nutrients are expressed as digestible, metabolisable or net energy. In most feed recommendations no specific advice is given about energy sources in the feed. However, plasma glucose levels exert a direct effect on the reproductive capacity of the sow. Via insulin, it directly affects oocyte maturation, glycogen deposition in the foetal piglet and thereby it also affects piglet vitality. During lactation glucose is an important precursor of lactose and therefore important for milk production and development of the offspring. Consequently, the sow may have a specific requirement for glucose as energy source both during gestation and lactation.

Glucose can originate from different dietary constituents. It may be absorbed as glucose after the digestion of starch, or it can be synthesized in the liver from gluconeogenic amino acids, glycerol, from dietary fats or from propionic acid which results from the fermentation of carbohydrates in the hindgut. Furthermore, some glucose may become available to the animal from glycogen stores in the liver and muscles. The amount of glucose available from 100 gram digestible nutrients is presented in Table 1. The amount available from protein is an estimation of the glucose production from dietary amino acids.

In this paper we will discuss the effect of glucose metabolism on sow performance during gestation and lactation.

Table 1. Estimated production of glucogenic energy

Nutrient	Grams glucose / 100 g digestible nutrient
Starch	111
Sugar	105
Fermentable NSP	20
Fat	11
Protein[1]	60

[1] 70% glucogenic amino acids: loss of urea.

2. Glucose and gestation

The sow has a low specific need for glucose. Pere (2001) concluded, based on arterio-venous differences, that glucose is an important energy source for both the uterus and the foetal piglets. Arterio-venous differences in the umbilical cord showed that besides glucose also lactate was taken up by the foetus (respectively 0.34 and 0.26 mMol/l), whereas only small amounts of fatty acids and glycerol were used. Fowden *et al.* (1997) and Pere (2001) calculated that the amount of substrate oxidized in the foetal piglet consisted for 75% of carbohydrates (35-40% glucose and 25-30% lactate), 10-15% amino acids and minor amounts of fatty acids, glycerol, ketoacids and acetate. Thus, glucose is an important energy source for the foetal pig and the level of glucose in the diet of the gestating sow might positively affect foetal development and consequently litter performance.

Effect of starch level during gestation on litter performance

Bikker *et al.* (2002) performed an experiment in which sows were fed iso-energetic feed (NE basis) during the entire gestation period (day 0 - day 110 of gestation), varying in the starch plus sugar content from 140 to 300 g/kg (Table 2). Daily intake of sugar plus starch ranged from 450 to 950 g. Both piglet birth weight and litter weight were lowest at the lowest starch intake levels. Similar observations have been made by Van der Aar *et al.* (1988), who observed a higher mean birth weight if the dietary starch levels were above 230 g/kg. Van der Peet *et al.* (2004) observed a 50 g higher birth weight if gestating sows received 895 g starch plus sugar in the gestation diet from day 85 of gestation until farrowing, in comparison to the groups that received 540 g per day.

Table 2. Influence of the glucose content in gestation diets on reproductive performance of sows (Bikker et al., 2002).

Starch + sugar content (g/kg) Nutrient (g/kg)	300 65 fat	230 95 fat	140 125 fat
Starch + sugar intake (g/d)	950	700	450
Sow BF gain (mm)	2.9[bc]	3.2[c]	2.5[ab]
Litter size	11.6[a]	12.4[b]	11.6[a]
Birth weight (g)[1]	1,528[b]	1,524[b]	1,489[ab]
Litter weight (kg)	17.6[ab]	18.3[b]	17.1[a]

[1] Litter size as covariate.
[abc] Different superscripts indicate significant differences.

In an experiment by Van der Aar and Cornelissen (1987), in which the sows were fed a diet with 35.5% starch from day 60 of gestation onwards, piglet survival during the first 14 days was significantly higher than piglet survival of sows that had received diets with 26% starch or less. This suggests that the higher dietary starch content improved piglet viability, either by a somewhat higher birth weight or a higher milk production. Also in other experiments performed at Schothorst Feed Research (Bikker et al., 2002; Van der Aar et al., 1988) a higher body gain and/or more back fat was found, if additional starch was fed during gestation. It was concluded that glucose, being a major energy source for foetal piglets, positively affects birth weight when added as starch and sugars to the diet of gestating sows. However, it is not clear whether the effect of glucose is present during entire gestation or if it is more pronounced during a specific period e.g. end of gestation.

3. Insulin resistance

Insulin resistance

During the last trimester of gestation, growth rate of foetuses is high and in order to reach optimal growth, the foetuses mainly depend on glucose and lactate as energy sources. In pigs, glucose transport from the sow to the foetus is dependent on the difference in blood glucose concentration between the sow and the foetus. Generally, the blood glucose level in the sow is at least 2.5 times higher than the blood glucose level of the foetus (Pere, 1995) in order to facilitate transport of glucose to the foetus. Lactate is formed from glucose in the swine placenta and the blood lactate level is, in contrast to glucose, 2.6 times higher in the foetal than the sow's blood (Pere, 1995). In order to increase glucose transport to the foetus it has been shown that many species, including the pig, develop a type of insulin resistance during the last trimester of gestation. This type of insulin resistance is reversible and reduces the rate of glucose utilization by maternal tissues by reducing the sensitivity to insulin and delaying insulin secretion (Pere et al., 2000). Consequently, blood glucose levels remain higher for a longer period of time and more glucose passes the placenta to the foetus. It has also been postulated that in a number of species, insulin resistance remains present during lactation in order to stimulate glucose transport to the udder for milk production (Pere and Etienne, 2007). In sows, levels of insulin and glucose were found to be at least one third higher in the week pre-farrowing (insulin: 6.6 µM/ml; glucose: 4.1 µM/ml) than during lactation (insulin: 3.3-4.2 µM/ml; glucose: 3.1 µM/ml; Revell et al., 1998). Furthermore, insulin resistance seems to decrease as lactation progresses as a longer glucose half-life was found at day 16 (22.5 min) than day 27 (18.8 min) of lactation

(Quesnel *et al.*, 2007) and was lowest post-weaning (10.8 min; Pere and Etienne, 2007). This indicates that in sows, insulin resistance gradually diminishes during lactation. There are, however, indications that this may depend on the metabolic status of the sow. In other animals with a high body weight loss, insulin resistance may remain post-weaning, which can influence reproductive performance.

In summary, insulin resistance develops in sows during gestation in order to improve glucose transfer to the foetuses. During increased insulin resistance, insulin sensitivity is low resulting in prolonged high blood glucose levels after feed intake.

Insulin resistance: a way to improve piglet survival?

Energy reserves in new born piglets are low compared to offspring of other mammals that develop insulin resistance during gestation. This could be a result of the relatively very mild insulin resistance in sows. As piglet viability and survival are highly dependent on their energy reserves (ref), it is important that these energy reserves, mainly present as liver and muscle glycogen, are increased. It has been found that glycogen increases between days 60 and 110 of gestation (Randall and L'Ecuyer, 1976; Okai *et al.*, 1978). As insulin resistance prolongs the post absorptive increase in blood glucose, it may be possible to increase glycogen storage in piglets by means of influencing insulin resistance in the sow. In the following paragraph, the possibilities of affecting insulin resistance via nutrition are discussed. Furthermore, the effects on piglet parameters including glycogen reserves are discussed in 'Insulin resistance and piglet parameters'.

Sow nutrition affects insulin resistance

Development of insulin resistance during progressing gestation is a normal process in many species. Possibly, this process can be influenced by nutrition. In their review, Storlien *et al.* (2000) stated that fat intake is linked to insulin resistance. Saturated fats increase insulin resistance, whilst poly unsaturated fatty acids (PUFAs) seem to reduce insulin resistance. In a study conducted at Schothorst Feed Research (Bikker *et al.*, 2007) effects of nutrition on insulin resistance were examined by feeding sows either a high unsaturated fat diet during or low fat, high starch diet in order to improve glycogen reserves in the piglets. A glucose tolerance test was performed at day 7, day 84 and day 112 of gestation to examine the effects on insulin resistance. As shown in Figure 1, no differences in glucose or insulin levels were found before start of the treatments at day 7 of gestation. At day 84 of gestation, glucose and insulin levels remained higher for a longer period of time in sows fed the low fat diet. These results show that a reduction in the dietary U/S ratio

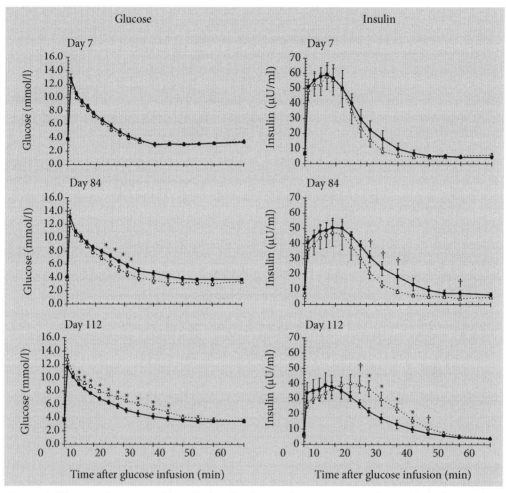

Figure 1. Plasma glucose and insulin levels of sows in response to glucose infusion at days 7, 84 and 112 of gestation fed either a control diet with an U/S ratio of 6.1 containing a high percentage of saturated fat and short chain unsaturated fatty acids (--Δ---) until lactation or a diet with a low U/S ratio of 1.5 containing 6% palm oil (—.—) until day 84 of gestation followed by a high starch diet (340 g/kg starch).
* indicates a significant (P<0.05) difference between the diets and † indicates a near significant trend (P<0.10).

from 6.1 to 1.5 can affect insulin resistance in gestating sows. A high fat diet, even a diet high in unsaturated fat, increases insulin resistance more than a low fat, high starch diet. Thus, the use of unsaturated fats can affect insulin resistance during gestation in sows. In contrast to fat, the role of protein in the development of insulin resistance is not very well studied. There are some indications in rats, that including L-glutamine in a high fat diet reduces insulin resistance (Storlien *et al.*, 2000).

In lactating sows, no effects of a high (19% CP) vs. a low (7.9%) protein diet were found on basal glucose and insulin levels. In this study, no glucose tolerance test was performed during lactation so it is not clear whether the protein level in lactation diets affects insulin resistance in lactating sows. In summary, nutrition can affect insulin resistance in sows. Studies indicate that the fatty profile may potentially increase insulin resistance, but these possible effects and also of other dietary sources such as protein need further investigation.

Insulin resistance and piglet parameters

As described in 'Insulin resistance: a way to improve piglet survival?', piglets have low energy reserves at birth and improvement of these energy reserves during fetal development may increase piglet survival. Nutrition of the gestating sow can improve glycogen development in the fetus. A number of studies have been performed in which the effects of inclusion of medium chain fatty acids in the gestation diet on piglet survival were studied.

In the study of Bikker *et al.* (2007) effects of the nutrition of the gestating sow on both piglet survival and sow parameters were studied. As described in 'Sow nutrition affects insulin resistance', during late gestation a high unsaturated fat diet resulted in a higher insulin resistance compared to a high starch diet. Piglets of these sows had increased liver glycogen at birth, However, this did not result in an increased viability but rather a higher mortality (mortality day 7 to weaning 0.48% vs. 3.28% in unsaturated fat diet). Survival of the piglets in the high strach treatment was not due to higher glycogen reserves or glucose levels at birth, but to glucose tolerance of the sows. Sows fed a high starch diet from day 85 of gestation onwards had a higher glucose tolerance and lower insulin resistance. Kemp *et al.* (1996) found a lower piglet survival during the first week post-weaning when glucose tolerance in sows was low (and thus insulin resistance high). Two possible explanations are given: (1) high insulin resistance results in higher levels of glucose for a longer period of time. Thus, fetuses are exposed to high glucose levels which are absent after birth resulting in a hypoglycemic shock, reducing survival; (2) low insulin resistance results in lower levels of glucose available for the fetuses which are then used to having low levels of glucose and are therefore better able to deal with low levels of glucose at birth. Based on this it was concluded that possibly not increasing insulin resistance during gestation but reducing insulin resistance improves piglet survival, which is not realized via increased glycogen levels. Insulin resistance on the other hand seems to improve the energy reserves of piglets at birth, but without improving piglet survival.

Feeding sows medium chain triglycerides (MCT) in gestation diets (about 10% of the diet) during the last three weeks of gestation significantly improved piglet survival from day 3 post-farrowing onwards compared to a control diet containing 19% starch and 2% soybean oil (Azain, 1993). Especially survival of the light piglets (birth weight <900 grams) to weaning was greater for piglets of which the sows were fed MCT (68%) compared to those of sows were fed the control diet (32%). Survival of the light piglets was also significantly improved when the sows were fed a diet with long-chain triglycerides (12% soybean oil). However, overall survival of piglets in this treatment was not significantly improved compared to the control diet (81.2% vs. 80.2%). Blood glucose was measured in the lighter third piglets at birth and significantly higher levels were found in piglets of which the sows were fed either MCT (76.8 mg/dl) or long-chain triglycerides (80.3 mg/dl) compared to the control diet (61.3 mg/dl). Jean and Chiang (1999) also found improved survival of lighter piglets (birth weigh <1,100 grams) during the first three days after birth when the sows were fed a diet containing coconut oil as MCT source (80.0%) or MCT (98.6%) from day 84 of gestation onwards compared to a control (C) diet with soybean oil (47.6%). In piglets of the sows fed either coconut oil or MCT, higher liver (coconut oil: 190.9 mg/g; MCT: 180.0 mg/g vs. C: 158.5 mg/g) and muscle glycogen levels (coconut oil: 75.8 mg/g; MCT: 84.6 mg/g vs. C: 65.4 mg/g) were found compared to control piglets. The outcomes of these studies indicate that piglet survival, especially of the lighter piglets within a litter, can be improved by nutrition of the gestating sow through improvement of the energy reserves of the piglets. MCT may improve energy reserves as the ketone bodies are formed out of the MCT pass the placenta and stimulate lipogenesis (Jean and Chiang, 1999). When lipogenesis in the foetus is stimulated, glucose is spared and can thus be stored as glycogen. However, in most of these studies the effects on sow blood glucose and insulin levels were not studied. Although it is known that fat intake is linked to insulin resistance, from these studies can not be concluded that piglet survival improved as a result of a MCT effect on insulin resistance in the gestating sow.

The effects of insulin resistance are contradictory. Insulin resistance increased glycogen levels in the piglets, but there are also indications that it has negative effects. More insight is needed in order to be able to optimally use the physiology of the sow in improving piglet survival.

4. Glucose and lactation

Glucose requirements of the lactating sow

Dietary glucose is the major precursor for lactose. Boyd *et al.* (1995) and Farmer *et al.* (2008) measured by means of arterio-venous differences, the uptake of nutrients by the mammary gland and the output of nutrients in the milk (Table 3). Both groups concluded that glucose was the major energy source as glucose accounted for 57 to 61% of the dry matter taken up by the mammary gland. Darragh and Moughan (1998), reported a dry matter content of sow milk of 18.7% and milk fat, protein and lactose contents of 7.6, 5.5 and 5.3% respectively. In different experiments at Schothorst Feed Research milk fat concentrations varying from 6.5 to 9.0%, milk protein from 4.5 to 5.5% and lactose from 5 to 5.8% were found.

Energy excreted as lactose as percentage of glucose energy uptake was 34% according to Boyd *et al.* (1995) and 43% according to Farmer *et al.* (2008). Energy as fat in milk was higher than triglycerides or fatty acids uptake. Energy uptake as amino acids was equal to energy excretion as protein. These data indicate that glucose is the major energy source for the mammary gland and *de novo* fat synthesis in the mammary gland is responsible for more than 50% of the milk fat content.

Furthermore, milk synthesis was found to be very energetically efficient. Both Boyd *et al.* (1995) and Farmer *et al.* (2008) measured an efficiency of 89%.

Based on these data, the two research groups estimated a glucose requirement of approximately 140 g/l milk produced. For an average

Table 3. Energetic efficiency of milk synthesis in the mammary gland of sows (Boyd et al., 1995; Farmer et al., 2008).

Precursor	Uptake (kJ/dl milk)		Product	Output (kJ/dl milk)	
	Boyd *et al.* (1995)	Farmer *et al.* (2008)		Boyd *et al.* (1995)	Farmer *et al.* (2008)
Glucose	220	213	Lactose	75	90
TAG/FA[1]	87	161	Fat	198	220
Amino acids	101	109	Protein	91	126
	408	488		364	436
Efficiency (%)				89	89

[1] TAG/FA: triaglycerols/fatty acids.

milk production of 12.5 l, this means that 1,750 g glucose is required per day. Information about maintenance requirement for glucose is limited. Linzell *et al.* (1969), estimated a glucose oxidation rate of 2.2 mg /min/kg body weight. This would equal a glucose amount of 750 g per day for a sow of 235 kg. Thus, depending on milk production a provisional estimate of glucose requirement of lactating sows would be approx. 2,500 g glucose per day.

Effect of nutrition on milk production and piglet performance

In several experiments at Schothorst Feed Research the effects of different energy sources on milk production and composition have been studied. Van der Aar and Cornelissen (1987), fed iso-energetic diets containing either 380 or 149 g starch/kg from day 60 of gestation till weaning. The diets did not result in differences in feed intake. At day 13 of lactation, the high starch diets resulted in a significantly higher milk production (11.1 vs. 10.2 kg/day), a higher lactose content (5.8 vs. 5.5%), but in a lower milk fat content (6.1 vs. 7.2%). The energy content in the milk per day was slightly lower in the high starch diet (4,718 vs. 4,797 kJ). High starch diets resulted in a higher output of lactose and protein in the milk per piglet and also piglet gain was higher. A high correlation was found between piglet growth and daily lactose plus protein consumption (r^2=0.96), indicating that lactose and protein are the major nutrients determining the growth of suckling piglets. The increased milk production could be explained by the higher glucose supply to the mammary gland on the high starch diets thus providing more glucose for lactose production in the udder. At the low starch diet amino acids may have been used for gluconeogenesis, thus rendering less amino acids available for milk protein synthesis, explaining a lower milk output. The effects of high starch in lactation diets on milk, milk protein production and piglet growth were also observed by other authors (Van den Brand, 2001; Bikker *et al.*, 2004). Van den Brand (2001) also showed that the sow body weight loss during lactation was lower when the sows were fed lactation diets high in starch.

In summary, the lactating sow requires high levels of glucose for milk production. Increasing the level of starch in the lactation diet resulted in a higher milk production with a higher lactose content, but lower fat and energy contents of milk. The growth rate of suckling piglets was improved. Thus, glucose during lactation is of importance and can positively affect sow and offspring performance.

Table 4. Effects of feeding dextrose either during lactation and weaning-to-oestrus interval (WOI) or during lactation on subsequent litter characteristics

	25 g dextrose and 25 g lactose/kg feed during lactation and WOI (Van den Brand *et al.*, 2009)			150 g dextrose/day during WOI (Van den Brand *et al.*, 2006)		
	No dextrose	Dextrose	SEM	No dextrose	Dextrose	SEM
Live born piglets	13.23	13.54	0.42	12.71	12.91	0.62
Birth weight (g)	1,483[a]	1,569[b]	30	1,591	1,608	49
CV birth weight (%)	22.6[a]	19.9[b]	0.9	21.2[a]	17.5[b]	1.3
Mortality birth to weaning (%)	13.35	12.06	-	7.4	6.9	-

[a,b] different superscripts indicate significant differences between 'dextrose' and 'no dextrose'.

5. Glucose and reproduction

Although the mechanisms are not yet fully understood, glucose may affect sow reproduction. Two recent studies have shown that supplementation of dextrose (25 g/kg diet) and lactose (25 g/kg diet) during the last week of gestation and lactation (Van den Brand *et al.*, 2009) or topdressing of dextrose (150 g/day) on feed from weaning to oestrus (Van den Brand *et al.*, 2006), significantly reduced the coefficient of variance of birth weight in both studies, and improved mean birth weight in one study (Table 4). The underlying mechanism of the effect of dextrose on within-litter variation in birth weight may possibly involves insulin. It is known that supplying easily available glucose as dextrose results in insulin peaks. A number of studies have shown that insulin affects reproductive parameters such as litter size (Ramirez *et al.*, 1997). It has been shown that exogenous administration of insulin can affect reproductive parameters in primiparous sows. For example, exogenous administration of insulin (0.4 IU/kg BW) once daily for four days starting on the day after weaning, significantly improved farrowing rate compared to sows treated with saline (92.3% vs. 76.7%; Ramirez *et al.*, 1997). Also, administering exogenous insulin for four days improved litter size (10.3±0.3) compared to no insulin administration (9.5±0.1) or two days of insulin administration (9.1±0.3; Ramirez *et al.*, 1997). The effect on litter size found may be realized by an improved follicle and early embryo development at higher insulin levels during the period around weaning, which can be influenced for example by the addition of glucose or dextrose to the diet. More research is needed to find out if glucose administration affects follicle development and early embryo development via insulin.

Tokach *et al.* (1992) found a correlation between insulin levels on day 7 during lactation and the number of luteinizing hormone (LH) peaks during lactation. Furthermore, insulin levels at day 21 of lactation tended to be higher in sows showing oestrus within 9 days post-weaning compared to sows showing oestrus after more than 15 days post-weaning (15.1 ± 0.9 µIU/ml vs. 10.6 ± 1.1 µIU/ml). However, in a number of other studies no correlations were found between insulin levels and LH secretion in primiparous sows (e.g. Van den Brand *et al.*, 2000). At present, it is not clear via which exact mechanism insulin in response to dietary glucose, acts on the reproductive system of the sow. However, it can be concluded that nutritional glucose can positively affect reproductive parameters of sows.

6. Conclusions

Increasing the level of available glucose for the sow via nutrition by increasing the level of starch in the gestation or lactation diets and increasing the level of dextrose during the weaning to oestrus interval, positively affects sow reproductive performance. During gestation, glucose is an important energy source for the uterus and the developing foetus. Increasing the level of starch in the gestating diet has been shown to increase piglet birth weight but not piglet vitality. During lactation, increasing the starch level in the diet positively affects milk production and consequently also improves piglet gain. The importance of glucose, especially at the end of gestation, is already demonstrated by the gestating sow as at the end of gestation sows develop insulin resistance by which more glucose is facilitated to the foetus. This mechanism could be of interest to improve piglet viability, but a number of studies have shown that piglet survival is decreased when insulin resistance is increased. This needs further investigation. Also the positive effects of glucose on follicle development during the weaning to oestrus interval needs to be studied as it is of interest to know if litter size and homogeneity of the new born litter can already be affected via nutrition during the weaning to oestrus interval.

In conclusion, glucose is an important source of energy during all stages of the sows' reproductive cycle. Further studies are needed to examine the optimal glucose levels in the diets for optimal sow performance and the mechanisms through which it affects sow reproduction need to be unravelled.

References

Azain, M.J., 1993. Effects of adding medium-chain triglycerides to sow diets during late gstation and early lactation on litter performance. J. Anim. Sci. 71: 3011-3019.

Bikker, P., Dirkzwager, A., Fledderus, J., 2002. De invloed van de energiebronnen zetmeel, vet en fermenteerbare koolhydraten op de technische resultaten, gedrag en wateropname van dragende zeugen. Schothorst Feed Research report 604.

Bikker, P., Dirkzwager, A., Fledderus, J., 2004. Glucosevoorziening van lacterende zeugen. Schothorst Feed Research report 640.

Bikker, P., Kluess, J., Fledderus, J., Geelen, M.J.H., 2007. Invloed van voeding van dragende zeugen op vitaliteit en glycogeenreserves van pasgeboren biggen. Schothorst Feed Research report 801.

Boyd, D.R., Kesinger, R.S., Harrell, R.J., Bauman, D.E., 1995. Nutrient uptake and endocrine regulation of milk synthesis by mammary tissue of lactating sows. J. Anim. Sci. 73: 36-56.

Darragh, A.J., Moughan, P.J., 1998. The amino acid composition of human milk corrected for amino acid digestibility. Br. J. Nutr. 80: 25-34.

Farmer, C., Trottier, N.L., Dourmad, J.Y., 2008. Review: Current knowledse on mammary blood flow, mammary uptake of energetic precursors and their effects on sow milk yield. Can. J. Anim. Sci. 88: 195-204.

Fowden, A.L., Forhead, A.J., Silver, M., MacDonald, A.A., 1997. Glucose, lactate and oxygen metabolism in the fetal pig during late gestation. Exp. Physiol. 82: 171-182.

Jean, K.-B, Chiang, S.-H., 1999. Increased survival of neonatal pigs by supplementing medium-chain triglycerides in late-gestating sow diets. Anim. Feed Sci. and Technol. 76: 241-250.

Kemp, B., Soede, N.M., Vesseur, P.C., Helmond, F.A., Spoorenberg, J.H., Frankena, K., 1996. Glucose tolerance of pregnant sows is related to postnatal pig mortality. J. Anim. Sci. 74: 879-885.

Linzell, J.L., Mepham, T.B., Annison, E.F., West, C.E., 1969. Mammary metabolism in lactating sows: arteriovenous differences of milk precursors and the mammary metabolism of [14C] glucose and [14C] acetate. Br. J. Nutr. 23: 319-332.

Okai, D.B., Wyllie, D., Aherne, F.X., Ewan, R.C., 1978. Glycogen reserves in the fetal and newborn pig. J. Anim. Sci. 46: 171-187.

Pere, M.-C., 1995. Maternal and fetal blood levels of glucose, lactate, fructose, and insulin in the conscious pig. J. Anim. Sci. 73: 2994-2999.

Pere, M.-C., 2001. Effects of meal intake on materno-foetal exchanges of energetic substrates in the pig. Reprod. Nutr. Dev. 41: 285-296.

Pere, M.-C., Etienne, M., 2007. Insulin sensitivity during pregnancy, lactation, and postweaning in primiparous gilts. J. Anim. Sci. 85: 101-110.

Pere, M.-C., Etienne, M., Dourmad, J.Y., 2000. Adaptations of glucose metabolism in multiparous sows: effects of pregnancy and feeding level. J. Anim. Sci. 78: 2933-2941.

Quesnel, H., Etienne, M., Pere, M.-C., 2007. Influence of litter size on metabolic status and reproductive axis in primiparous sows. J. Anim. Sci. 85: 118-128.

Randall, G.C.B., L'Ecuyer, C.L., 1976. Tissue glycogen and blood glucose and fructose levels in the pig fetus during the second half of gestation. Biol. Neonate. 28: 74-82.

Ramirez, J.L., Cox, N.M., Moore, A.B., 1997. Influence of exogenous insulin before breeding on conception rate and litter size of sows. J. Anim. Sci. 75: 1893-1898.

Revell, D.K., Williams, I.H., Mullan, B.P., Ranford, J.L., Smith, R.J., 1998. Body composition at farrowing and nutrition during lactation affect the performance of primiparous sows: I. Voluntary feed intake, weight loss, and plasma metabolites. J. Anim. Sci. 76: 1729-1737.

Storlien, L.H., Higgins, J.A., Thomas, T.C., Brown, M.A., Wang, H.Q., Huang, X.F., Else, P.L., 2000. Diet composition and insulin action in animal models. Brit. J. Nutr. 83, suppl. 1: S85-S90.

Tokach, M.D., Pettigrew, J.E., Dial, G.D., Wheaton, J.E., Crooker, B.A., Johnston, L.J., 1992. Characterization of luteinizing hormone secretion in the primiparous, lactating sow: relationship to blood metabolites and return-to-estrus interval. J. Anim. Sci. 70: 2195-2201.

Van den Brand, H., 2001. Energy partitioning and reproduction in primiparous sows: Effects of dietary energy source. PhD Thesis Wageningen University, Wageningen, The Netherlands.

Van den Brand, H., Dieleman, S.J., Soede, N.M., Kemp, B., 2000. Dietary energy source at two feeding levels during lactation of primiparous sows: I. Effects on glucose, insulin, and luteinizing hormone and on follicle development, weaning-to-oestrus interval, and ovulation rate. J. Anim. Sci. 78: 396-404.

Van den Brand, H., Soede, N.M., Kemp, B., 2006. Supplementation of dextrose to the diet during the weaning to estrus interval affects subsequent variation in within-litter piglet birth weight. Anim. Reprod. Sci. 91: 353-358.

Van den Brand, H., Van Enckevort, L.C.M., Van der Hoeven, E.M., Kemp, B., 2009. Effects of dextrose plus lactose in the sows diet on subsequent reproductive performance and within litter birth weight variation. Reprod. Dom. Anim. Doi 10.111/j.1439-0531.2008.01106.

Van der Aar, P.J., Cornelissen, J.P., 1987. De voeropname bij jonge zeugen. Schothorst Feed Research report 225.

Van der Aar, P.J., Veen, W.A.G., Cornelissen, J.P., 1988. De invloed van de voeding tijdens de dracht op de voeropname tijdens de zoogperiode en op de melkvet- en huidvetsamenstelling. Schothorst Feed Research report 237.

Van der Peet, C.M.C., Kemp, B., Binnendijk, G.P., Den Hartog, L.A., Vereijkens, P.F.G., Verstegen, M.W.A., 2004. Effects of additional starch or fat in late-gestating high nonstarch polysaccharide diets on litter performance and glucose tolerance in sows. J. Anim. Sci. 82: 2964-2971.

Sustainable pig house: first results of slurry flushing and the CYCLIZ® process

Christian Leroux[1], David Guillou[2] and Bernard Raynaud[3]
[1]INZO, Centre de Recherches Zootechniques Appliquées, les Simons, 02540 Montfaucon, France
[2]INZO, 1 rue de la Marébaudière, Montgermont, B.P. 96669 35766 saint Grégoire Cedex, France
[3]Union InVivo, Division Agriculture Durable, 83 Av de la Grande Armée, 75782 Paris Cedex 15, France; braynaud@invivo-group.com

Abstract

Slurry flushing under the pigs leads to improve the indoor conditions of the breeding. A large quantity of water is necessary to run this technique. Cycliz®, a patented physicochemical process for manure treatment, enables to recycle the water needed for the flushing. The process generates a high pH effluent. The antiseptic properties of this effluent improve the hygienic condition and the health of the pigs. The flush twice to four times a day get better employees' working conditions as well as zootechnical performance. In the effluent out of the station, compared to the initial slurry, the treatment reduces by 91.7% the nitrogen quantity. Moreover, it enables to keep nearly all the phosphorus in organic fertilizer and a large amount of the nitrogen as mineral fertilizer.

1. The pig breeding in France along with sustainable development

The concentration of pig farming, mainly in the West of France, has leaded to the creation of many regulations in order to manage, and then reduce the environmental impact of the pig breeding. For instance, the French ICPE (Environment Classified Installations) regulates the creation or the extension of pig farms, and controls the pollution rates, while the nitrate directive strictly regulates the nitrogen and phosphorus pressure/hectare, and the spreading period of the slurry in the fields.

In addition, a complementary regulation is being introduced in terms of ammonia and potassium emissions.

To counterbalance these nitrogen excesses, more than 300 manure treatment plants were built in the last 15 years in the West part of France. They process 10 to 15% of the manure in this area. These plants were built, at first, as a mean to reduce the quantity of nitrogen spread on the farm fields.

These biological stations, through a nitrification-denitrification process, transform manure nitrogen into gaseous nitrogen (N_2). To comply with an additional regulation on phosphorus, these stations are more and more upstream joined with a centrifugation or a sieving operation, which enables to separate the liquid and solid parts of the manure. The solid part concentrates most of the phosphorus. It is sent to a composting station and after maturing, is exported out of the structural excess areas.

The technical and economical performances of these stations are very high. The nitrogen reduction rate is approximately 60%, while the phosphorus reduction rate induced by the compost export is 80% and an additional nitrogen 20%. On an economical point, the direct cost of treatment is limited, below 3 €/slurry m^3. This includes energetic and maintenance costs. So beyond the initial investment subsides, these stations operate without any public grant.

2. InVivo's sustainable agriculture strategy

The Union of cooperatives InVivo, leading agricultural company in Europe, centralises both purchasing, sales and services. The main activities are: seeds, agricultural supplies, storage, grain trading, animal nutrition and health, suburban retail. The turnover was 5,278 million Euros in 2008, a fourth on animal feed activities.

The mission of InVivo is to design, develop and share techniques, products, tools and services that allow cooperatives and farmers to produce more and better.

Plant division

InVivo offers cooperatives and farmers decision-making software designed to improve crop system profitability. These software are made to optimize fertilisation and crop protection programs. They also deliver information on the best spraying time, based on the computerised development of pests or on the nitrogen needs of crops. Each year, these software are used on a very large acreage. For instance for fertilization, Epicles, is used on more than 1,500,000 ha each year by roughly 700 coop advisers.

From these registered informations, it's possible to assess 16 environmental impact indicators (Table 1). They will be included within the sofwares in order to enable the agronomists to deliver at the same time agronomic and environmental advices. This evaluation of environmental consequences is major information for the farmers and

Table 1. Environmental impact indicators.

Item	Pressure	Management	State

Water quality	Nitrogen rate
	Phosphorus rate
	Pesticides:
	• frequency treatment
		Nitrogen balance
		Phosphorus balance
		(Nitrate) in leaching
		  waters
		(Pesticides) in leaching
		  and running waters
		Winter soil coverage
			Quality of the mass
			  water:
			• Nitrate
			• Pesticides
			• IBGN

Climate	Green house gas
	  emissions
		Green house gas
		  balance
			Average temperature

Energy	Fossil energy • quantity used	Crop energy balance (Bio fuel)	

their counsellors, so that they can define and implement productive production strategy and efficient environmental management.

Moreover, during the large French consultation on sustainable development named the 'Grenelle de l'environnement', the environmental labelling of the food products based on their production processing and distribution chain has been selected. So, informations coming from such agronomic and environmental databases could, eventually, be used to promote the environmental performance of cultural systems.

Animal division

InVivo has a research station in order to test and improve the feed range offered to the farmers. It enables to anticipate many regulatory constraints imposed to the pig breeding: two-phase diet, antibiotics withdrawal from feed… However, nuisances linked to this activity stop its development. They affect neighbours discomfort (smell), environment (nitrogen and phosphorus pressure, emission of greenhouse gases – NH_3). For instance the lower efficiency of manure effluent when used as nitrogen fertilizer for crops, generates an alteration of the nitrogen gross balance in comparison with mineral fertilizer performance. So the nitrogen leaching risks during winter could be increased. In the same way, concentration in heavy metals of the composted products or potash of the slurry treatment effluent implies a good management of the spreading amount to avoid alteration of the agronomical potential of the soil.

Finally, in the rooms, ammoniac air pollution in an excessive concentration or dusts… lead to discomfort or even lung diseases of the farm workers. Antibiotics are frequently needed in pig farms to treat the animal diarrhoea, intestinal and lung disorders.

Since 1999, the work of the research station has been progressively adapted to test different farming processes and manure treatments to solve these disagreements. All these tests led to the flushing Cycliz® process.

3. CRZA pig research station

The CRZA (Centre de Recherches Zootechniques Appliquées) is located 10 km from Château Thierry in the Aisne department in France. This centre covers an area of 22 hectares. The experimental unit works with the main farm animals bred for production in France: cattle, sheep, swine, poultry and rabbits.

Reproduction

The CRZA farm is a farrow to finish unit with 105 productive sows. Its production capacity is 2,500 marketed pigs per year. It has 800 pig places in the fattening section and 504 in the post-weaning section. Herd productivity was 28.5 weaned piglets per sow per year in 2008, i.e. 2,756 piglets produced per year; the surplus is sold to fatteners.

The building is adapted to an all in – all out management. The reproduction unit is made of 7 rooms: 2 breeding rooms, 2 gestation rooms, 2 farrowing rooms with 12 places and a buffer area with 6 places. The different rooms are managed on an all in – all out basis.

Postweaning

The CRZA has three 128-places rooms. Each room consists of 32 pens for 4 piglets. The pens have a slatted cast steel/plastic floor. The pits are mid-depth (60 cm) and fitted with PVC gutters for daily flushing. Room climate is monitored through ventilation, with a central extracting fan which allows the control of air-speed ('Exavent') (capacity: 5,400 m³/h). In this unit, feeding is ad libitum.

Fattening

The CRZA has three types of rooms in the fattening section:
- 4 rooms for liquid feeding and group housing including two rooms equipped with flushing systems. Each room consists of 16 pens for 8 pigs. Two independent lines for liquid feeding are operated in the four rooms, with capacity for delivering any pen indvidually (one electronic valve per pen).

- One room with IVOG individual feeding stations for group-housed pigs. This room is also fitted with a flushing system.
- Two rooms for individual housing without any flushing system.

In each of these rooms the pigs are weighed individually. Body composition could be approached using Real Time Ultrasound scanning on farm or through the carcass grading measurements performed at the slaughterhouse.

Ventilation is managed by air pumping under the slats 11,000 m^3/h.

In the rooms equipped with liquid feeding, feed is always allowed according to a restricted feeding schedule, usually with a plateau after the 5th or 6th week in the room.

Flushing experimentation

This experimentation was conducted at the CRZA in parallel feeding trials. The groups in the flushed houses and the groups with manure under the pigs received the same diet. These tests were performed first in post weaning, then in fattening stages.

The slurry coming from all the rooms is collected and mixed in a reception tank and treated then with the Cycliz® process. Flushes were realized with the liquid effluent recycled from this treatment station. The pH of this effluent is 12.

4. The CYCLIZ® process

Pig rooms equipment

In order to save water and ensure good faeces elimination, gutters are layed under the slatted floor. They enable to concentrate material and ensure an efficient flush, without any preferential way for the water. This installation ensures a total elimination of faeces with a limited amount of liquid.

During post weaning 1 or 2 flushing operations are done each day depending on piglet age. On average, 320 litres are used per flushing and per room; that is 2,5 l/piglet/flush.

During fattening: depending on pig age 2, 3 then 4 flushing operations are done each day, 800 litres are used per flush; that is 6,25 l/pig/flush.

Flush cycles are managed by a programmable robot.

The slurry treatment station Cycliz® and the co-products

The process is described in Table 2. The station was installed in 2005. It has been running nonstop since January 2008. The process leads

Table 2. Operation diagram of Cycliz® process.

Raw manure input to station				
	% treated	**Volume m³**	**Nitrogen kg N**	**Phosphorus kg P_2O_5**
Raw manure	100%	**1000**	**2,270** *2.27*	**950** *0.95*
Additive related to process		53	0	0
Ouptput product				
Reduction on liquid phase (including recycling into the flushing)	**36.4%**		**91.7%**	**~100%**
Cake	2% 21		4.2% **95.55** *4.55*	7.4% **70.56** *3.36*
Cycliz liquid	90.8% 956		11.8% **267.68** *0.28*	0% **traces** *<0.002*
Centrifugation ouptut solid	5.7% 60		32.7% **743.4** *12.39*	88.7% **842.4** *14.04*
Ammonium sulphate	1.5% 16		48.7% **1,105.6** *69.1*	0% **0** *0*

to the effluent, which is of zootechnical interest and agronomic related products (Figure 1):

- the centrifugation residue, which is easily composted, contains 32.5% of nitrogen and most of the phosphorus. This product can be exported to other areas. It's a substitute to synthetic phosphate fertiliser;
- solid from filter press: this product can be used as soil liming material;
- nearly 50% of the manure nitrogen is collected within the ammonium sulphate solution. After a pH rectification, this product can be used as any mineral fertiliser. Moreover, it brings sulphur fertilisation, very useful for oilseed rapes or cereals. Considering its low nitrogen level, 7% ammonium sulphate solution, this product must be used in a reasonable distance, no more than a hundred kilometres away from the pig farm;

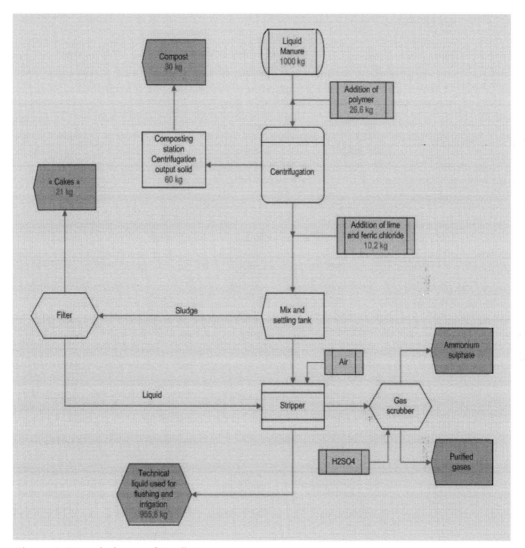

Figure 1. Mass balance of Cycliz® process.

• a third of the effluent is used for the flush. Its very high pH guarantees its microbiological quality. The unused effluent is kept on the exploitation before spreading. It contains most of the slurry potash.

All these products do not present any olfactory disagreement. Furthermore, ammoniac emanation coming from the storage and the centrifugation are collected and sent to the gas scrubber with the air coming from the stripping engine.

Environmental performances

The mass balance is described in Figure 1. If all the compost and the ammonium sulphate solution are exported, the reduction rates in nitrogen, phosphorus and micro material in water suspension are respectively 87, 100 and 99% (Figure 2).

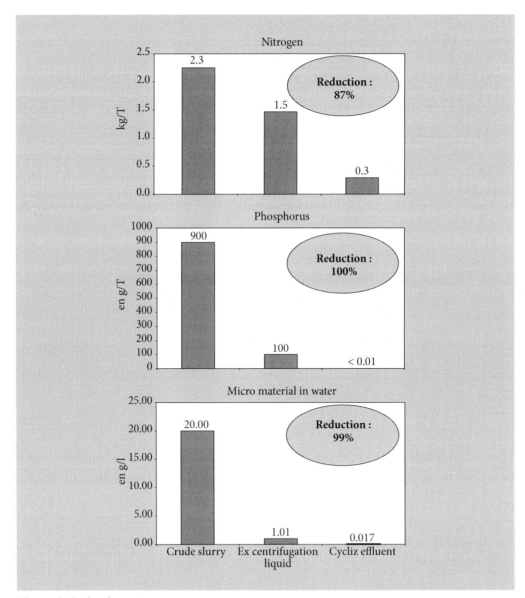

Figure 2. Reduction rates.

The Cycliz® process enables the farmer to keep on the farm the necessary quantity of fertilizers for the crops. These fertilizers have a very efficient form which enables easier management of the environmental impact, as they are be brought just in time and quantity needed by the crops.

Air quality in the pig house

- Ammonia: controls have been done during 15 visits in one month. The measure were done with a portable gas monitoring device (Oldham MX 2100). The precision of which was 1 ppm. The results are showed in Figure 3. The average level in flushed rooms was reduced by five fold compared to unflushed.
- Methane: according to Pablo Meisner and Herman Van den Weghe (LANDTECHNIK 04/03), methane emissions are reduced up to 90%.
- Volatile compounds: a study was conducted at the CRZA by ARD in 2006, in the postweaning room with group number 321 (with flushing) and 322 (without flushing) and on an empty room used as a control. The air coming from the three rooms were submitted both to two different analysis.
- A sensory test by a panel formed of 87 persons (administrative personnel, technical staff from the station, external visitors from the swine industry, other visitors).

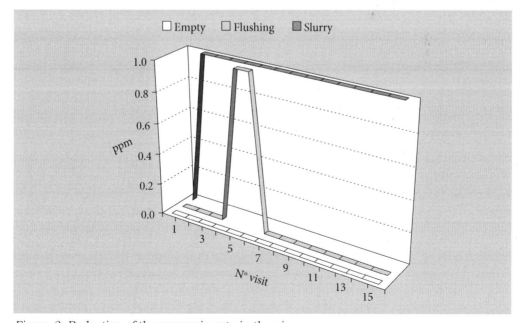

Figure 3. Reduction of the ammonia rate in the air.

In the room with pigs and no flushing, 82% of the responses ranked the smell 'offensive' or higher, whereas less than 20% of responses declared the smell offensive in the flushed room (Figure 4).

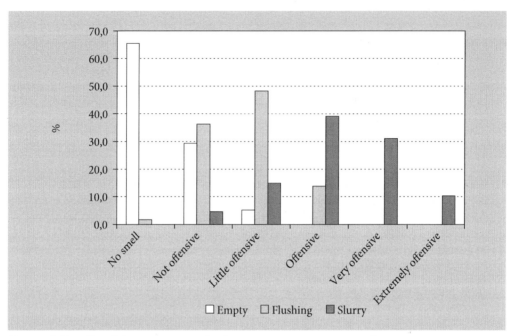

Figure 4. Results of the sensory analysis.

Performances in postweaning

At the CRZA, the piglets are weaned at 28 days and they don't receive any regular treatments against diarrhoea, nor any alternative antibiotic products in food. On the 96 piglets only 9 were injected with antibiotic in unflushed rooms compared to 5 in the flushed ones (Table 3):
- the respiratory and digestive problems (diarrhoea) were significantly reduced in the flushed rooms;
- mortality rate was reduced to zero;
- the percentage of soft faeces was the same in both rooms. However, liquid faeces were detected only in the non flushed room and had to be cured;
- the average daily gain (ADG) presented a 5.8% increase (Table 4) whereas the food conversion ration (FCR) remained constant;
- these results are correlated to the fact that, during the first two weeks, the feed intake of the flushed piglets rised, with no digestive problems (Table 5).

Table 3. Zootechnical performances in postweaning.

	Flushing		Slurry	
Weaned piglets	96		96	
Dead	0		2	2%
Wasting	3	3%	2	2%
Health cares				
Leg troubles	2	2%	3	3%
Respiratory	0		4	4%
Diarrhea	0		1	1%
Fecal scores				
Soft	3	13%	3	13%
Liquid	0		1	4%
Number of vetenarian treatments	9/96		5/96	

Table 4. Zootechnical performances in postweaning.

Post weaning	Slurry stored under slats	Slurry cycliz flushing
Entering weight (kg) 0 day	8.7	8.2
Exit weight (kg) 41 days	32.4	33.3
ADG (g/day) 0 - 41 days	579	613
Global FCR	1.46	1.47

Table 5. Zootechnical performances in postweaning.

	Flushing		Slurry		
	Mean	±SE	Mean	±SE	
Phase 1 ADG	422	12.5	369	11.6	$P<0.001$
Phase 1 ADFI % BWw	4.1%	0.12%	3.5%	0.10%	$P<0.001$
Phase 2 ADG	722	10.0	699	10.8	$P<0.05$
Global FCR	1.47	0.014	1.46	0.014	N.S.

Performances in fattening

Test was been performed on group number 371 and 372 in flushed rooms compared to group 370 in a non flushed room. Within each group, pens of 7-9 pigs were constituted based on initial body weight with female pigs and castrated males, ending with 8 blocks of 2 pens per weight category. In fattening, feed was offered restricted. The feed quantity was 1,500 g on day 1 and increased by 25 g/day. It was then limited to a maximum amount: 2,600 g/day till market weight. The same feed was given to the three batches; genetic origin of the pigs was the same.

The results in Table 6 show:
- a 6.1% increase of the ADG and a FCR of 2.47;
- a better flexibility at the beginning fattening stages: under the CRZA conditions, the diet is optimum when introducing a 26-30 kg piglet in fattening, as shown in diagram 3 for the non flushed pigs. In the flushed room, the FCR during this period is lower, and this performance is independent from piglet's weight. On the contrary, during the second fattening period, the feed restriction erased the differences between the two batches of pigs (Figure 5 and 6).

Table 6. Performances in fattening.

Fattening	Slurry stored under slats	Slurry cycliz flushing
Entering weight (kg)	30	29
Exit weight (kg)	114.8	116.8
ADG (g/day)	834	885
Fattening times (days)	102	100
Global FCR	2.62	2.47

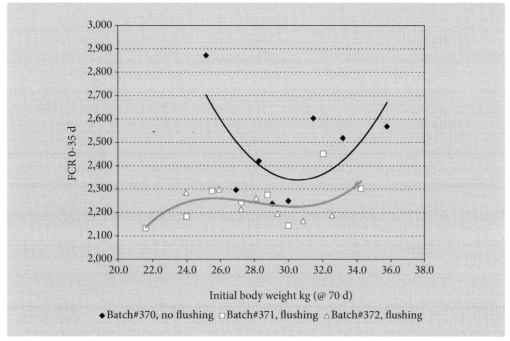

Figure 5. First 5 fattening weeks – Feed conversion ratio.

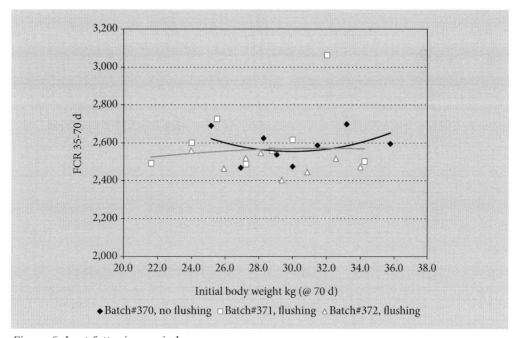

Figure 6. Last fattening period.

5. Discussion

Flushing the manure under the pigs, several times a day, with an antiseptic liquid, enhances considerably the employees' working conditions. Moreover, it enables to improve sanitary conditions and related zootechnical performances. On the other hand, this flushing increases by 30% the volume of slurry to treat and requires to recycle the flushing water.

The patented process Cycliz® produces an antiseptic liquid which reinforces the sanitary security of the breeding. Besides, it enables to keep 50% of the nitrogen as mineral fertilizer and nearly all the phosphorus.

Operating costs, due to necessary inputs, are higher than those in well known biological stations. But the net costs of the process are supposed to be equal. The zootechnical advantages observed in the high sanitary conditions and strict cleanness of the CRZA research station are expected to be at least the same but probably higher under practical farm conditions. Furthermore, because of the higher sanitary conditions in the pig houses using the Cycliz® process feed costs can be reduced either through reduced medication costs or the of less complex feeds.

All these improvements, combined with the co products valorisation and the savings on veterinary treatments, should lead to develop 'green' pig houses, both highly efficient and without damages for the neighbourhood and the environment.

Energy in poultry diets: adjusted AME or net energy

Jan Dirk van der Klis[1], Cees Kwakernaak[1], Alfons Jansman[2] and Machiel Blok[3]
[1]*Schothorst Feed Research, P.O. Box 533, 8200 AM Lelystad, The Netherlands; jdvdklis@schothorst nl*
[2]*Wageningen UR, Animal Sciences Group. P.O. Box 65, 8200 AB Lelystad, the Netherlands*
[3]*Centraal Veevoeder Bureau (CVB), P.O. Box 29739, 2502 LS The Hague, the Netherlands*

Abstract

Energy evaluation in poultry is generally based on metabolisable energy systems. Since many decades it is questioned whether development of a net energy system would have added value to predict energy partitioning in meat- and egg-producing poultry. Based on the extra-caloric value of fat in laying hens and a potential lower energetic efficiency of protein utilisation, several modified ME systems were developed and used in commercial poultry nutrition. Recently, CVB initiated the development of an ATP-based NE system for broilers. Validation of this energy system did not show added value over the current modified AME system for broilers. Therefore, most probably frequent updates in nutrient digestibilities in feedstuffs would be of higher commercial relevance, as well as target animal dedicated digestibility data.

1. Introduction

Feed costs are the main costs for poultry production, of which 70 to 75% are related to dietary energy. To be able to reduce feed costs while maintaining bird performance, dietary energy values should be an accurate representation of the utilisable energy for both meat- and egg-producing birds in their successive production phases. This not only refers to the energy evaluation system used, but it should be realised that digestibilities of the energy delivering nutrients varies among bird species and age. Generally, energy evaluation systems in poultry are based on metabolisable energy (apparent metabolisable energy: AME, or true matabolisable energy: TME), being calculated from digestible protein, digestible fat and digestible carbohydrates. This carbohydrate fraction might be split into starch, sugars and non starch polysaccharides.

Discussions are ongoing whether adjustment of AME or TME systems would have added value for poultry, taking the efficiency of utilisation of digested nutrients into account. Does it improve the predictability of bird performance and/or reduce feed costs? Furthermore, the potential of a net energy (NE) system in poultry analogous to swine needs to be evaluated, as it has been shown in swine nutrition that feed cost savings can be as high as 4.00 to 4.50 €/metric ton when a shift is made from a ME to a NE system, without any negative impact on production performance. The latter might even be improved due to lower dietary crude protein levels, which are generally linked to the introduction of a NE system. Of course feed cost savings depend on the feedstuffs available in different regions in the world and on the maximum dietary feedstuff inclusion levels for different animal species or categories. In principle, a NE system would have added value in poultry like in pigs, however, results in poultry are less conclusive. Differences between ME and NE in pigs are based on a higher efficiency of utilisation of energy from fat and a lower efficiency of utilisation of energy from protein compared to starch. In laying hens the extra caloric value of fat has been demonstrated (eg. Scheele *et al.*, 1985 as cited by Van der Klis and Fledderus, 2007) and is well-accepted, but results in broilers are still contradictory. Nevertheless, in time several modified AME systems have been developed that take the efficiency of nutrient utilisation into account, like
- Effective Energy by Emmans (1994).
- For egg-producing poultry using a 15% extra caloric value of digested fat and for meat-type poultry reducing the energy value of digested protein by approx. 15% (both modifications being used in commercial poultry feeding in the Netherlands (CVB, 2005)).

Recently, the CVB in the Netherlands initiated the development of an ATP-based NE system for broilers. The added value of this new system over the currently used AME_broiler was evaluated. In 2009, Schothorst Feed Research launched a new feeding table for laying hens, based on digestibility experiments with highly productive layers. In the current paper results are presented.

2. Modified AME system in laying hen diets

The basis of all energy evaluation systems lies in accurate nutrient digestibilities in feedstuffs. Until now, feeding values determined in adult roosters are used world-wide to calculate the dietary energy value for highly productive layer diets. Recently, Schothorst Feed Research determined nutrient digestibilities and AME values using white LSL laying hens in peak production according to the European reference assay for poultry (Bourdillon *et al.*, 1990), based on a three-day droppings collection period. Nineteen feedstuffs were evaluated so far and results were compared with current tabulated values from restrictedly fed adult roosters (CVB, 2005). Results are given in Figure 1.

It is shown in Figure 1 that especially the digestibility of carbohydrates was higher in *ad libitum* fed laying hens compared to restrictedly fed roosters. Differences in nutrient digestibilities resulted in absolute AME values in feedstuffs (as measured) that were -40 to 530 kcal/kg higher than current tabulated values (CVB, 2005), whereas the feeding value of fats was in general overestimated in the current feeding tables. Although test feedstuffs in Figure 1 were not simultaneously evaluated in restrictedly fed roosters, it was concluded based on the large differences between laying hens and adult roosters that the latter have limited value to determine the energy value of feedstuffs for producing laying hens nowadays. Schothorst Feed Research therefore introduced in 2009 a new feedstuff table for layers.

In the Netherlands a modified AME system for laying hens is being used, taking a higher efficiency of energy utilisation for fat compared to

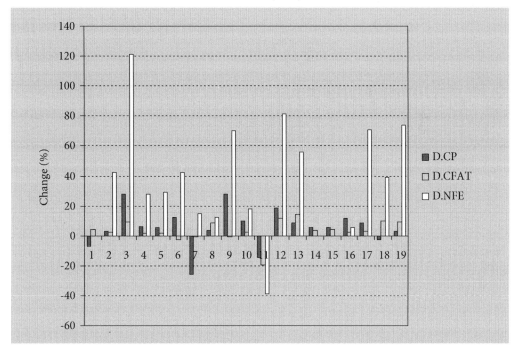

Figure 1. Nutrient digestibilities (digestible crude protein (D.CP), digestible crude fat (D.CFAT) and digestible nitrogen free extract (D.NFE)) in feedstuffs using laying hens in peak production (Schothorst Feed Research, 2009). Data are presented relative to current tabulated values (i.e. restrictedly fed adult roosters) as a reference (CVB, 2005).

starch into account, as fatty acids and monoglycerides can be used as such to synthesize fat in body and eggs. Scheele *et al.* (1985) estimated the energetic efficiency of fat utilisation in comparison to starch using *ad libitum* fed laying hens from 24 to 36 weeks of age. Two sets of diets contained either 11.5 or 12.0 MJ AME_N/kg (AME corercted for zero nitrogen balance), each with three levels of dietary fat (3, 6 and 9%) that was isocalorically exchanged for carbohydrates and ash. Dietary AME, nutrient digestibilities and energy retention in body and eggs were measured. Results are given in Figure 2.

It is shown in Figure 2 that energy retention in laying hens increases with isocaloric dietary fat inclusion for starch, indicating a higher energetic efficiency of fat utilisation over carbohydrates. From this experiment it was calculated that the extra-caloric value of fat for layers was 15% to 19%. The CVB uses a 15% extra caloric value of fat in the Dutch modified AME system for layers to take this higher efficiency into account. In laying hens no adjustment has been made for a lower efficiency of protein utilisation, except for the energy correction for zero nitrogen balance, assuming that all nitrogen from feed will be excreted as uric acid.

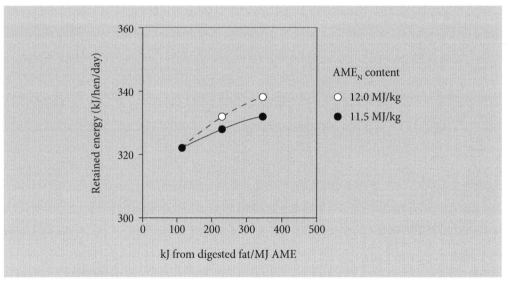

Figure 2. The energy retention (in body and egg) in laying hens fed diets differing in AME_N content, each containing three levels of dietary fat which were isocalorically exchanged for carbohydrates (Scheele et al.*, 1985).*

3. Development of an ATP-based net energy system

In 2004 a NE formulae was derived based on the ATP yield of carbohydrates, amino acids, glycerol and fatty acids and volatile fatty acids (Jansman *et al.*, 2004). The results of their calculations of the ATP yield of nutrients is given as mol ATP/g nutrient and recalculated to kJ/g in Table 1.

From these coefficients the following equation to calculate the NE_{ATP} in feedstuffs was derived:

$$NE_{ATP} = 9.7 \times \text{dig. true protein} + 26.1 \times \text{dig. crude fat} +$$
$$+ 11.7 \times \text{dig. starch} + 10.6 \times \text{dig. sugars} + \qquad (1)$$
$$+ 8.2 \times \text{dig. nitrogen free residue}$$

Subsequently, an equation to calculate the NE_{ATP} requirement (kJ/day) was derived from comparative slaughter experiments based on regression analyses between the NE_{ATP} intake and protein and fat deposition, using male and female broilers from ten different pure and commercial lines (unpublished data), being:

$$NE_{ATP \, (req)} = 278 \text{ kJ } NE_{ATP}/BW^{0.75}/\text{day} \times BW^{0.75} +$$
$$+ 3.058 \times \text{energy in protein deposition (kJ } NE_{ATP}/\text{day)} + \quad (2)$$
$$+ 1.053 \times \text{energy in fat deposition (kJ } NE_{ATP}/\text{day)}$$

Based on Equation 1 the NE_{ATP} content of feedstuffs has been calculated and compared to the AME content, using maize as a reference (Table 2). It is suggested in Table 2 that the relative NE_{ATP} of protein rich feedstuffs will be similar to the currently used $AME_{broiler}$.

Table 1. The ATP yield and ATP energy per g nutrient (absolute and relative to starch).

	mol ATP/g	ATP energy (kJ/g)	Relative to starch
Glucose	0.211	10.56	90
Starch	0.235	11.73	100
Protein[1]	0.194	9.70	83
Fat[2]	0.508	25.31	216
Volatile fatty acids	0.212	10.60	90
Fermentable carbohydrates	0.165	8.21	70

[1] Based on the composition of body protein.
[2] Based on soya oil.

Table 2. Relative feeding values of vegetable feedstuffs, using corn as a reference.

	$AME_{broiler}$[1] (growing)	AME poultry (adult)	NE_{ATP}
Maize	100	100	100
Wheat	91	94	84
Soybean meal	59	67	58
Sunflower seed meal	43	48	42

[1]AME broiler is based on an approx. 15% lowered energy value of digestible protein.

The NE_{ATP} system has been validated in Ross 308 broilers based on a set of five diets with a fixed $AME_{broiler}$ (12.2 MJ/kg) and variable NE_{ATP} (7.7 to 8.3 MJ/kg) and five diets with a fixed NE_{ATP} (8.0 MJ/kg) and variable $AME_{broiler}$ (11.8 to 12.6 MJ/kg) to test whether the new NE_{ATP} equation has added value over the current $AME_{broiler}$ equation to predict the energy retention in broilers. Broilers were fed twice a day from day 8 onwards, and daily feed allowance was standardised based on pairwise feeding. Nutrient digestibilities were measured from 13-17 days of age and from 27-31 days of age and energy retention was determined using the comparative slaughter technique from 8 to 24 days of age and from 24 to 36 days of age. Average feed intake from 8 to 36 days was 82.7 g/day (approx. 75% of *ad libitum* intake for Ross 308 in this period). Body weight gain during the experimental period was 1,557 g, resulting in an average feed conversion efficiency of 1.491.

The energy retention from 8 to 36 days of age is given in Figure 3. AME and NE were related in this experiment, because the current modified $AME_{broiler}$ equation was used. Approx. 65.0% of the variation in energy deposition was explained by $AME_{broiler}$ intake and only 60.3% by NE_{ATP} intake. It was therefore concluded that in this experiment NE_{ATP} had no added value over $AME_{broiler}$ (Jansman and Van Diepen, 2008), in which broilers were restrictedly fed.

4. Net energy versus metabolisable energy system

Pirgozliev and Rose (1999) quantitatively reviewed the relationship between dietary AME, NE and Effective Energy on the one hand and energy deposition (or net energy for production: NEp) on the other hand, using a rather old dataset of 40 feedstuffs varying in AME value from 8.0 to 16.5 MJ/kg. NEp was measured as energy deposition in young growing chickens depositing mostly lean meat (Fraps 1946, as cited by

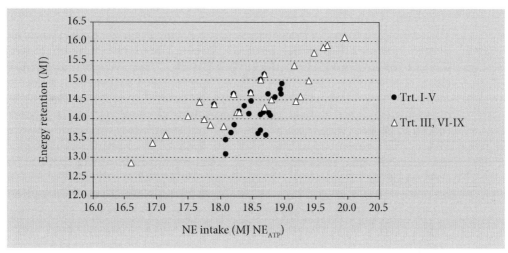

Figure 3. The Energy retention (MJ) in broilers from 8 to 36 days of age as affected by NE_{ATP} intake. Diets I to V were formulated on variable $AME_{broiler}$ content (keeping NE_{ATP} as constant as possible) and diet III and VI to IX on variable NE_{ATP} content (both in MJ/kg).

Pirgozliev and Rose, 1999). Based on regression analyses, Pirgozliev and Rose (1999) concluded that the AME value (calculated as 22.4 × D.CP+ 39.2 × D.CFAT + 17.2 D.NFE (in MJ/kg)) overestimated NEp for high protein feedstuffs. The regression equations obtained (after we omitted two feedstuffs (dried butter milk and dried skimmed milk) with a high leverage from their dataset) were:

NEp= 0.711 × AME – 0.598 (r^2=0.92) for cereal and cereal by-products and
NEp= 0.545 × AME + 0.843 (r^2=0.81) for high protein feedstuffs.

These authors concluded that correcting the AME for crude protein content gave an equally accurate prediction of NEp as Effective Energy according to Emmans (1994) and Net Energy according to Hoffmann and Schiemann (1980) and De Groote (1974) as cited by Pirgozliev and Rose (1999). This is more or less as expected as the dietary AME value was not yet corrected for zero nitrogen balance, using an energy value of 22.4 instead of 18.0 MJ/kg digested CP. Based on their regression analyses, the energy value of digested protein most probably needs to be corrected below 18.0 MJ/kg D.CP. Pirgozliev and Rose (1999) did not show any proof for an extra caloric value of fat in broilers, which can be attributed to the lack of feedstuffs with higher fat levels in their dataset. Nitsan *et al.* (1997) did show an extra caloric effect of 2.6% soybean oil (plus wheat bran) isocalorically exchanged for maize meal in broilers at 12.1 (comparing 2.6 and 5.3% fat) and 13.1 MJ/kg (comparing 6.9 to

9.5% fat). At the low dietary AME level heat production was reduced by 14% due to soybean oil inclusion, whereas effects at the high AME level were only small. Based on their calculations the extra caloric effect was 17% in the low, and 3% in the high AME diet. They suggested that the extra caloric effect is related to bird age (no effect expected in young birds having a low fat deposition), dietary fat level and source. An excess of unsaturated fatty acids might even stimulate heat production. More recently, Noblet *et al.* (2009) were not able to show any effect of the dietary fat content in broiler diets on their heat production. The NE/ME ratio in their trial was approx. 75% irrespective of the dietary fat content (2.9% versus 9.6%). Noblet *et al.* (2007) did not demonstrate any increase in heat production either using high protein diets in 4-5 week old broilers. The NE/ME ratio was remained 68% although the dietary crude protein level was increased from 22.5 to 27.3%. It can therefore be questioned whether a net energy system would have any significance to broilers, despite a few references that might indicate some benefits of a NE system. A (modified) AME system most probably will be adequate to describe energy partitioning in broilers and preferable above a more complex NE system.

References

Bourdillon, A., Carré, B., Conan, L., Francesch, M., Fuentes, M., Huyghebaert, G., Janssen, W.M.M.A., Leclerq, B., Lessire, M., McNab, J., Rigoni, M., Wiseman, J., 1990. Br. Poultry Sci. 31: 567-676.

Emmans, G.C., 1994. Br. J. Nutr. 71: 801-821.

Jansman, A.J.M., Kwakernaak, C., Van der Klis, J.D., De Bree, J., Fledderus, J., Brandsma, G.G., Blok, M.C., 2004. A net energy system for broilers (in Dutch). ASG-report 04/5917.

Jansman, A.J.M., Van Diepen, J.T.M., 2008. Validation of a new net energy system for broilers (in Dutch). ASG-report 08/.

Nitsan, Z., Dvorin, A., Zoref, Z., Mokady, S., 1997. Br. Poultry Sci. 38: 101-106.

Noblet, J., Dubois, S., Van Milgen, J., Warpechowski, M., Carré, B., 2007. Heat production in broilers is not affcted by dietray crude protein. In: Ortigues-Marty, I., Miraux, N. Brand-Williams, W. (eds.) Energy and production metabolism and nutrition. Wageningen Academic Publishers, Wageningen, the Netherlands, pp. 479-480.

Noblet, J., Warpechowski, M., Dubois, S., Van Milgen, J., Carré, B., 2009. Influence of feed fat content on metabolic utilization of energy in growing chickens. In: Proc. Of the 8e Journées de la Recherche Avicole, Saint-Malo, France.

Pirgozliev, V., Rose, S.P., 1999.WPSA Journal 55: 23-36.

Scheele, C.W., Van Schagen, P.J.W., Van Es, A.J.H., 1985. Energy utilisation in layer diets (in Dutch). COVP report 005.

Van der Klis, J.D., Fledderus, J., 2007. Evaluation of raw materials for poultry: What's up? In: Proc. of the 16th European Symposium on Poultry Nutrition, Strasburg, France.

Molting in laying hens and *Salmonella* infection

Steven C. Ricke[1], Claudia S. Dunkley[2], Jackson L. McReynolds[3], Kingsley D. Dunkley[4] and David J. Nisbet[3]
[1]University of Arkansas, Center for Food Safety and Department of Food Science, 2650 N. Young Ave. Fayetteville AR 72704-5690, USA; sricke@uark.edu
[2]University of Georgia, Department of Poultry Science, P.O. Box 748, Tifton, GA 31793, USA
[3]U.S. Department of Agriculture, Agricultural Research Service, Southern Plains Agricultural Research Center, 2881 F&B Road, College Station, Texas 77845, USA
[4]Abraham Baldwin Agricultural College, 2802 Moore Highway, Tifton, GA 31793, USA

Abstract

Molting is a natural occurrence in the avian species. It involves the shedding and replacement of feathers and the regression of the reproductive system thereby giving the birds a rest from reproduction. Reproductive molt is of particular interest to the commercial U.S. layer industry as it involves a period of declining productivity leading to reduced profitability for egg producers. Historically, complete removal of feed was practiced by the industry due to the fact that it was economical and delivered productive additional egg laying cycles. However, feed withdrawal can lead to the depression of the hens' immune system and changes in the gastrointestinal (GI) microflora, making laying hens susceptible to *Salmonella* infection. To alleviate these concerns, the development of alternative molting has become a focus of recent research efforts. Several approaches have been examined based on their ability to reduce physiological susceptibility in the hen and retain fermentative GI microflora. However, a key requirement for these approaches is their ability to effectively induce a molt and bring about post-molt performance that is similar to that of feed deprived hens.

1. Introduction

Beginning in the 1970's and through the 1990's *Salmonella enterica* serovar Enteritidis (SE) came to prominence as a leading cause of human foodborne salomonellosis (Guard-Petter, 2001; Schroeder *et al.*, 2006). Symptoms include fever and diarrhea typically associated with foodborne illness but in some cases can result in much more serious illness and death (Schroeder *et al.*, 2006). Early reports linked SE outbreaks to consumption of contaminated table eggs and this association with table eggs sparked interest in how laying hens become infected with this particular *Salmonella* and produced contaminated eggs (Gantois *et al.*, 2009; Gast, 1994; Golden *et al.*, 2008; Holt, 2003; Ricke, 2003; Schroeder *et al.*, 2006). Gantois *et al.* (2009) noted that SE contamination of shell eggs can occur either by internal penetration of the egg shell or internal contamination of the egg contents by infected reproductive organs during formation of the egg but which route is primary is still debated. Although serotypes such as *S.* Braenderup and *S.* Kentucky have been isolated from layer flocks (Kretzschmar-McCluskey *et al.* 2008; Li *et al.*, 2007) there are believed to be unique characteristics associated with the SE serotype and host conditions that make the laying hen particularly vulnerable to this serotype; these have been reviewed elsewhere and will not be discussed here (Dunkley *et al.*, 2009; Gantois *et al.*, 2009; Golden *et al.*, 2008; Guard-Petter, 2001). In the current review the primary focus will be on the specific molting practices that were identified as key factors in dissemination of SE in laying hen flocks.

2. Molting in the U.S. commercial egg industry

Domestic hens like most species of wild birds experience a naturally occurring molt. This however, is usually incomplete and hens continue to lay eggs at low rates for a prolonged period of time (Berry, 2003). This indicates the end of the useful life of a flock, which for a one-cycle (non-molted) program hens is approximately 80 weeks old (Bell, 2003). Over the years, researchers have developed methods of artificial molt induction to occur at times other than at the time of natural molting. It was in the 1930's that egg producers in the U.S. first adapted the practice of induced molting (Anderson and Havenstein, 2007). Egg producers at that time used multiple-molting programs in which hens were molted more than once and the laying cycles were shortened (30 weeks) with the first molt occurring on average at approximately 60 weeks and the second molt being induced at approximately 100 weeks (Bell, 2003). Californian producers were the first to extensively adopt the practice (Bell, 2003) which spread across the U.S. to the most major egg producing regions in the mid 1970's. By the early 2000's Bell (2003) stated that more than 75% of the commercial laying facilities in the U.S. implement the practice of induced molting to rejuvenate their flocks.

There are three basic ways by which a molt is induced; feed removal or limitation, low-nutrient ration, and feed additives (Park *et al.*, 2004c). Each of these methods typically involves the alteration of the photoperiod from a long day to a short day with the end result being a cessation of lay and the loss of feathers. This is done to give the egg producer maximum production from hens on the commercial layer farms by enabling them to have a second or even a third laying cycle potentially extending the useful reproductive life from less than 80 weeks to more than 110 weeks or even 140 weeks (Bell, 2003). Bell (2003) concluded that there was improved post-molt performance in molted hens compared to their non-molted counterparts with peak egg production occurring during the second cycle at approximately 77 to 79% 10 to 20 weeks into the second cycle. Anderson and Havenstein (2007) compared white and brown-egg layers that had been maintained in continuous production without molting versus birds artificially molted and concluded molted birds yielded nearly one dollar more per bird on average than their nonmolted counterparts.

3. Feed withdrawal molt induction

Historically, commercial egg producers in the U. S. typically used feed removal as the method to induce a molt (Park *et al.*, 2004c). Conventional feed withdrawal programs involved the removal of feed for a period of 5 to 14 days, water (no more than 3 days) or both from the hens and also reducing the photo-period in the houses to a natural day length or less to cause complete involution of the reproductive tract (Bell, 2003; Berry, 2003; Park *et al.,* 2004c). A wide range of physiological and behavior studies have compared nonmolted with molted birds in an attempt to dissect out responses specific to birds undergoing molt. Dunkley *et al.* (2008a,b) video recorded and quantified a variety of hen behaviors in 10 minute intervals over a 9 day molt period. Not surprisingly feeder activity decreased pecking in birds undergoing feed withdrawal while nonnutritive pecking increased toward the end of the feed withdrawal period. This coincides with the earlier report by Dunkley *et al.* (2007b) of an increase in a stress indicator serum protein during the latter stages of feed withdrawal. Metabolic shifts also occur during feed withdrawal molt including changes in serum cholesterol, ketone bodies, uric acid and triglceride levels during the molt period that reflect physiological adjustments to loss in body weight and reduction of the reproductive organs (Dunkley *et al.*, 2007a; Landers *et al.*, 2007, 2008). Likewise, decreases in serum calcium are indicative of the rapid cessation of eggshell formation and changes in bone structure and bone composition (Kim *et al.*, 2006, 2007, 2008).

4. Molting and *S. enteritidis* infection in laying hens

Feed withdrawal causes increased susceptibility to SE infection which is usually marked by increased intestinal shedding and colonization in internal organs such as the liver, oviducts, spleen, and ovaries (Gast, 1994; Holt, 2003; Ricke, 2003). The two primary reasons these birds become susceptible to SE infection are reductions in protective gastrointestinal microflora and reduced immune response capabilities (Dunkley *et al.*, 2009; Holt, 2003; Ricke, 2003). The role of the immune system has been previously summarized by Holt (2003) and will not be discussed here.

In general the microbial composition and fermentation activities in the chicken gastrointestinal tract are sufficiently complex to support extensive fermentation of a wide variety of feedstuffs *in vitro* (Dunkley *et al.*, 2007c; Saenkerdsub *et al.* 2006) *In vivo* fermentation acids are generally in greatest concentration in the ceca and cecal anaerobic bacteria have been consistently identified in young chicks and older birds including methanogens with both cultural and molecular methods (Józefiak *et al.*, 2004; Saenkerdsub *et al.* 2007a,b). It has been reported that feed withdrawal alters the microenvironment of the crop and ceca in the intestines of the fowl (Dunkley *et al.*, 2007d; Durant *et al.*, 1999; Ricke, 2003). Corrier *et al.* (1997) observed that though induced molting had no apparent effect on the pH or oxidation reduction potential of the ceca, it caused a reduction in acetic acid, propionic acid and total volatile fatty acids (VFA) in the ceca. This shift in cecal VFA has also been reported by others (Dunkley *et al.*, 2007d; Ricke *et al.*, 2004b, Woodward *et al.* 2005). Durant *et al.* (1999) observed increased crop pH, and reduced *Lactobacilli* populations and VFA concentrations in the crops of feed deprived hens. Molecular-based studies employing denatured gradient gel electrophoresis (DGGE) have revealed detectable shifts in cecal microbial populations with feed withdrawal birds when compared with birds continually fed (Dunkley *et al.*, 2007d; Ricke *et al.*, 2004a). Dunkley *et al.* (2007d) demonstrated that DGGE analysis of fecal samples could be used to estimate changes in microbial diversity during the entire molt induction period and could potentially be used as a noninvasive means to monitor changes in cecal microflora.

The SE intestinal shed rate is higher in birds that are exposed to exogenous sources of the pathogen concomitantly during molt induction. These hens had a more severe infection when compared to nonmolted hens (Holt *et al.*, 1994, 1995). Problems associated with SE in the flock environment can increase when birds are exposed to stresses such as feed deprivation and the spread of SE in larger numbers of susceptible hens in a flock is of major importance. The impact of air-borne transmission must also be taken into consideration since transmission to birds in cages which were a distance away from infected birds during a molt by feed withdrawal, could still be detected (Holt *et al.*, 1998). Air sampling may represent an opportunity to implement molecular methods for rapid detection of low levels of SE at early stages of flock infestation before the pathogen becomes widespread in the flock (Kwon *et al.*, 2000; Park *et al.*, 2008).

Durant *et al.*, (1999, 2000) demonstrated increases in the SE virulence regulatory gene, *hilA* when SE was incubated in the presence of crop contents from feed withdrawal laying hens. This increase in virulence expression was further demonstrated *in vivo* when Dunkley *et al.* (2007e) detected nearly a doubling in SE *hilA* levels in fecal and cecal contents on days 6 and 12 of the feed withdrawal molt period. Depending on the study SE can be recovered in the albumin and yolks of eggs from hens for several weeks after they have been orally inoculated (Gast, 1994). This is believed to be evidence for supporting that internal contamination of eggs can occur prior to the laying of the egg (Gast, 1994). In summarizing the current literature Gantois *et al.* (2009) concluded that SE possesses genes that allow it to survive the antimicrobial microbial factors associated with the albumen and vitelline membrane prior to residing in the yolk.

5. Control measures

There are several potential approaches that offer opportunities for limiting SE dissemination in laying flocks and reducing risk of appearance on shell eggs. Some focus has been directed toward either the initial phases of egg production or in the later retail stages. Vaccination of hens and rapid detection in feeds at the beginning stages of layer flock production such as the breeder flocks represent possible vertically – based preventative approaches to limit SE (Golden *et al.*, 2008; Jarquin *et al.*, 2009). At the terminal end of table egg production Schroeder *et al.*, (2006) proposed that rapid cooling and pasteurization of the eggs themselves would provide a highly effective means to limit risk of illness from SE contaminated eggs.

Most of the research has been focused on limiting horizontal SE dissemination in flocks during molting. Much of this research has been directed towards developing alternative methods to feed withdrawal for inducing molt. In addition to limiting SE, economical advantages to the commercial farmers by successfully providing a second laying cycle of high production rates and high quality eggs were identified as important criteria (Ricke, 2003). Historically, attempts have been made to incorporate nutritional imbalances, ingredients that decrease appetite, or addition of fillers to induce molts (Bell, 2003; Berry, 2003; Park *et al.*, 2004c) as means of inducing a molt. Lowering calcium content either directly or by increasing the zinc to calcium ratio has been the most extensively examined (Park *et al.*, 2004c). Both low and high zinc dietary levels have been shown to reduce SE establishment (Moore *et al.*, 2004; Ricke *et al.* 2004b) but inconsistent molt induction, repression of feed consumption and potential environmental emissions are potential problems for practical application (Park *et al.*, 2004a,b,c).

Reducing energy content either by restricted feed intake or addition of dietary components such as plant extracts to initiate molt have also been explored over the past several years (Park *et al.*, 2004c). Plant and cereal grain byproducts examined for molt induction potential have included alfalfa, barley, cotton seed, grape pomace, rice hulls, tomaoto extract, wheat gluten and middlings (Asadi Khosholi *et al.*, 2006; Biggs *et al.*, 2003; Landers *et al.*, 2005a,b; Mansoori *et al.*, 2007; Onbaşilar and Erol, 2007; Soe *et al.*, 2007, 2009). Reduction of SE in laying hens with wheat middlings and alfalfa – based diets has been demonstrated (Seo *et al.*, 2001; Woodward *et al.*, 2005). However, inconsistent SE reductions in hens molted with 100% alfalfa have also been observed and may be related to several factors associated with variable feed intake including initial behavior responses as well as passage rate effects (Dunkley *et al.*, 2008a,b,c; Woodward *et al.*, 2005). Feed particle size may also need to be taken into account when high fiber diets are being fed (Hetland *et al.*, 2005; Huang *et al.*, 2006). There is some indication that consistency can also be improved by either reducing the total level of alfalfa in the molt diet and/or combining it with other additives such as prebiotics, biological extracts or combining with nonfeed interventions (Donalson *et al.*, 2005, 2007, 2008a,b; McReynolds *et al.*, 2005, 2006; Willis *et al.*, 2008).

6. Conclusions

Alternative methods of inducing a molt that do not involve feed withdrawal continue to be the focus of research and several alternatives have been examined. While some of the alternative diets that involve nutrient restriction or dietary additives have been successful in inducing molts, they often are expensive, hard to adapt or are not readily available in all egg producing geographical regions. Although diets such as wheat middlings, and alfalfa appear to be effective in inducing a molt developing more predictable dietary modification strategies will require a better understanding of the GI microflora responses to dietary shifts and how these shifts directly influence SE colonization and invasion in the laying hen. Further evaluation of physiological, immunological and behavioral responses of molting hens on alternative diets as well as commercial merits will be essential in validating their potential importance as alternative molt diets to the table egg industry.

Acknowledgements

This research was supported by Hatch grant H8311 administered by Texas Agricultural Experiment Station, USDA-NRI grant # 2002-0614 and U.S. Poultry and Egg Association grant #485.

References

Anderson, K.E., Havenstein, G.B., 2007. Effects of alternative molting programs and population on layer performance: Results of the thirty-fifth North Carolina layer performance and management test. Journal of Applied Poultry Research 16: 365-380.

Asadi Khoshoei, E., Khajali, F., 2006. Alternative induced-molting methods for continuous feed withdrawal and their influence on postmolt performance of laying hens. International Journal of Poultry Science 5: 47-50.

Bell, D.D., 2003. Historical and current molting practices in the U.S. table egg industry. Poultry Science 82: 965-970.

Berry, W.D., 2003. The physiology of induced molting. Poultry Science 82: 971-980.

Biggs, P.E., Douglas, M.W., Koelkebeck, K.W., Parsons, C.M., 2003. Evaluation of nonfeed removal methods for molting programs. Poultry Science 82: 749-753.

Corrier, D.E., Nisbet, D.J., Hargis, B.M., Holt, P.S., DeLoach, J.R., 1997. Provision of lactose to molting hens enhances resistance to *Salmonella enteritidis* colonization. Journal of Food Protection 60: 10-15.

Donalson, L.M., Kim, W.K., Chalova, V.I., Herrera, P., McReynolds, J.L., Gotcheva, V.G., Vidanović, D., Woodward, C.L., Kubena, L.F., Nisbet, D.J., Ricke, S.C., 2008a. In vitro fermentation response of laying hen cecal bacteria to combinations of fructooligosaccharide prebiotics with alfalfa or a layer ration. Poultry Science 87: 1263-1275.

Donalson, L.M., McReynolds, J.L., Kim, W.K., Chalova, V.I., Woodward, C.L. Kubena, L.F., Nisbet, D.J., Ricke, S.C., 2008b. The influence of a fructooligosacharide prebiotic combined with alfalfa molt diets on the gastrointestinal tract fermentation, *Salmonella* Enteritidis infection and intestinal shedding in laying hens. Poultry Sci. 87: 1253-1262.

Donalson, L.M., Kim, W.K., Chalova, V.I., Herrera, P., Woodward, C.L., McReynolds, J.L., Kubena, L.F., Nisbet, D.J., Ricke, S.C., 2007. *In vitro* anaerobic incubation of *Salmonella enterica* serotype *Typhimurium* and laying hen cecal bacteria in poultry feed substrates and a fructooligosaccharide prebiotic. Anaerobe 13: 208-214.

Donalson, L.M., Kim, W.K., Woodward, C.L., Hererra, P., Kubena, L.F., Nisbet, D.J., Ricke, S.C., 2005. Utilizing different ratios of alfalfa and layer ration for molt induction and performance in commercial laying hens. Poultry Science 84: 362-369.

Dunkley, C.S., Friend, T.H., McReynolds, J.L., Kim, W.K., Dunkley, K.D., Kubena, L.F., Nisbet, D.J., Ricke, S.C. 2008a. Behavior of laying hens on alfalfa crumble molt diets. Poultry Science 87: 815-822.

Dunkley, C.S., Friend, T. H., McReynolds, J.L., Woodward, C.L., Kim, W.K., Dunkley, K.D., Kubena, L.F., Nisbet, D.J., Ricke, S.C., 2008b. Behavioral responses of laying hens to different alfalfa-layer ration combinations fed during molting. Poultry Science 87: 1005-1011.

Dunkley, C.S., Kim, W.-K., James, W.D., Ellis, W.C., McReynolds, J.L., Kubena, L.F., Nisbet, D.J., Ricke, S.C., 2008c. Passage rates in poultry digestion using stable isotope markers and INAA. Journal of Radioanalytical and Nuclear Chemistry 276: 35-39.

Dunkley, C.S., McReynolds, J.L., Dunkley, K.D., Kubena, L.F., Nisbet, D.J., Ricke, S.C., 2007a. Molting in *Salmonella* Enteritidis-challenged laying hens fed alfalfa crumbles. III. Blood plasma metabolite response. Poultry Science 86: 2492-2501.

Dunkley, C.S., McReynolds, J.L., Dunkley, K.D., Njongmeta, L.N., Berghman, L.R., Kubena, L.F., Nisbet, D.J., Ricke, S.C., 2007b. Molting in *Salmonella* Enteritidis-challenged laying hens fed alfalfa crumbles. IV. Immune and stress protein response. Poultry Science 86: 2502-2508.

Dunkley, K.D., Callaway, T.R., Chalova, V.I., McReynolds, J.L., Hume, M.E., Dunkley, C.S., Kubena, L.F. Nisbet, D.J., Ricke, S.C., 2009. Foodborne *Salmonella* ecology in the avian gastrointestinal tract. Anaerobe 15: 26-35.

Dunkley, K.D., Dunkley, C.S., Njongmeta, N.L., Callaway, T.R., Hume, M.E., Kubena, L.F., Nisbet, D.J., Ricke, S.C., 2007c. Comparison of in vitro fermentation and molecular microbial profiles of high-fiber feed substrates incubated with chicken cecal inocula. Poultry Science 86: 801-810.

Dunkley, K.D., McReynolds, J.L., Hume, M.E., Dunkley, C.S., Callaway, T.R., Kubena, L.F., Nisbet, D.J., Ricke, S.C., 2007d. Molting in *Salmonella* Enteritidis challenged laying hens fed alfalfa crumbles II. Fermentation and microbial ecology response. Poultry Science 86: 2101-2109.

Dunkley, K.D., McReynolds, J.L., Hume, M.E., Dunkley, C.S., Callaway, T.R., Kubena, L.F., Nisbet, D.J., Ricke, S.C., 2007e. Molting in *Salmonella* Enteritidis challenged laying hens fed alfalfa crumbles I. *Salmonella* Enteritidis colonization and virulence gene *hilA* response. Poultry Science 86: 1633-1639.

Durant, J.A., Corrier, D.E., Stanker, L.H., Ricke, S.C., 2000. Expression of the *hilA* *Salmonella typhimurium* gene in a poultry *S. enteritidis* isolate in response to lactate and nutrients. Journal of Applied Microbiology 89: 63-69.

Durant, J.A., Corrier, D.E., Byrd, J.A., Stanker, L.H., Ricke, S.C., 1999. Feed deprivation affects crop environment and modulates *Salmonella enteritidis* colonization and invasion of leghorn hens. Applied and Environmental Microbiology 65: 1919-1923.

Gantois, I. Ducatelle, R., Pasmans, F., Hasesebrouck, F., Gast, R., Humphrey, T.J., Van Immerseel, F., 2009. Mechanisms of egg contamination by *Salmonella* Enteritidis. Federation of European Microbiological Societies Microbial Reviews 33: 718-738.

Gast, R.K., 1994. Understanding *Salmonella enteritidis* in laying chickens: the contributions of experimental infections. International Journal of Food Microbiology 21: 107-116.

Golden, N.J., Marks, H.H., Coleman, M.E., Schroeder, C.M., Bauer Jr., N.E., Schlosser, N.D., 2008. Review of induced molting by feed removal and contamination of eggs with *Salmonella enterica* serovar Enteritidis. Veterinary Microbiology. 131: 215-228.

Guard-Petter, J., 2001. The chicken, the egg and *Salmonella enteritidis*. Environmental Microbiology 3: 421-430.

Hetland, H., Svihus, B., Choct, M., 2005. Role of insoluble fiber on gizzard activity in layers. Journal of Applied Poultry Research 14: 38-46.

Holt, P.S., 2003. Molting and *Salmonella enterica* serovar Enteritidis infection: The problem and some solutions. Poultry Science 82: 1008-1010.

Holt, P.S., Mitchell, B.W., Gast, R.K., 1998. Airborne horizontal transmission of Salmonella enteritidis in molted laying chickens. Avian Diseases 42: 45-52.

Holt, P. S., Macri, N. P., Porter, Jr., R. E., 1995. Microbiological analysis of the early Salmonella enteritidis infection in molted and unmolted hens. Avian Diseases 39: 55-63.

Holt, P. S., Buhr, R.J., Cunningham, D.L., Porter, Jr., R.E., 1994. Effects of two different molting procedures on a *Salmonella enteritidis* infection. Poultry Science 73: 1267-1275.

Huang, D.S., Li, D.F., Xing, J.J., Ma, Y.X., Li, Z.J., Lv, S.Q., 2006. Effects of feed particle size and feed form on survival of *Salmonella typhimurium* in the alimentary tract and cecal *S. typhimurium* reduction in growing broilers. Poultry Science 85: 831-836.

Jarquin, R., Hanning, I., Ahn, S., Ricke, S.C., 2009. Development of rapid detection and genetic characterization of *Salmonella* in poultry breeder feeds. Sensors 9: 5308-5323.

Józefiak, D., Rutkowski, A., Martin, S.A., 2004. Carbohydrate fermentation in the avian ceca: A review. Animal Feed Science and Technology 113: 1-15.

Kim, W.K., Herfel, T.M., Dunkley, C.S., Hester, P.Y., Crenshaw, T.D., Ricke, S.C., 2008. The effects of alfalfa-based molt diets on skeletal integrity of white Leghorns. Poultry Science 87: 2178-2185.

Kim, W.K., Donalson, L.M., Bloomfield, S.A., Hogan, H.A., Kubena, L.F., Nisbet, D.J., Ricke, S.C., 2007. Molt performance and bone density of cortical, medullary, and cancellous bone in laying hens during feed restriction of alfalfa -based feed molt. Poultry Science 86: 1821-1830.

Kim, W.K., Donalson, L.M., Mitchell, A.D., Kubena, L.F., Nisbet, D.J., Ricke, S.C., 2006. Effects of alfalfa and fructooligosaccharide on molting parameters and bone qualities using dual energy X-ray absorptiometry and conventional bone assays. Poultry Science 85: 15-20.

Kretzschmar-McCluskey, V., Curtis, P.A., Anderson, K.E., Kerth, L.K., Berry, W.D., 2008. Influence of hen age and molting treatments on shell egg exterior, interior, and contents microflora and *Salmonella* prevalence during a second production cycle. Poultry Science 87: 2146-2151.

Kwon, Y.M., Woodward, C.L., Pillai, S.D., Peña, J., Corrier, D.E., Byrd, J.A., Ricke, S.C., 2000. Litter and aerosol sampling of chicken houses for rapid detection of *Salmonella typhimurium* contamination using gene amplification. Journal of Industrial Microbiology & Biotechnology 24: 379-382.

Landers, K.L., Moore, R.W., Dunkley, C.S., Herrera, P., Kim, W.K., Landers, D.A., Howard, Z.R., McReynolds, J.L., Byrd, J.A., Kubena, L.F., Nisbet, D.J., Ricke, S.C., 2007. Immunological cell and serum metabolite response of 60 week old commercial laying hens to an alfalfa meal molt diet. Bioresource Technology 99: 604-608.

Landers, K.L., Moore, R.W., Herrera, P., Landers, D.A., Howard, Z.R., McReynolds, J.L., Byrd, J.A, Kubena, L.F., Nisbet, D.J., Ricke, S.C., 2008. Organ weight and serum triglyceride responses of older (80 week) commercial laying hens fed an alfalfa meal molt diet. Bioresource Technology 99: 6692-6696.

Landers, K.L., Howard, Z.R., Woodward, C.L., Birkhold, S.G., Ricke, S.C., 2005a. Potential of alfalfa as an alternative molt induction diet for laying hens: Egg quality and consumer acceptability. Bioresource Technology 96: 907-911.

Landers, K.L., Woodward, C., Li, X., Kubena, L.F., Nisbet, D.J., Ricke, S.C., 2005b. Alfalfa as a single dietary source for molt induction in laying hens. Bioresource Technology 96: 565-570.

Li, X., Payne, J.B. Santos, F.B., Levine, J.F. Anderson, K.E., Sheldon, B.W., 2007. *Salmonella* populations and prevalence in layer feces from commercial high-rise houses and characterization of the *Salmonella* isolates by serotyping, antibiotic resistance analysis, and pulsed field gel electrophoresis. Poultry Science 86: 591-597.

Mansoori, B., Modirsanei, M., Farkhoy, M., Kiaei, M.-M., Honarzad, J., 2007. The influence of different single dietary sources on moult induction in laying hens. Journal of the Science of Food and Agriculture. 87: 2555-2559.

McReynolds, J.L., Moore, R.W., Kubena, L.F., Byrd, J.A., Woodward, C.L., Nisbet, D.J., Ricke, S.C., 2006. Effect of various combinations of alfalfa and standard layer diet on susceptibility of laying hens to *Salmonella* Enteritidis during forced molt. Poultry Science 85: 1123-1128.

McReynolds, J., Kubena, L., Byrd, J., Anderson, R., Ricke, S., Nisbet, D., 2005. Evaluation of *Salmonella enteritidis* in molting hens after administration of an experimental chlorate product (for nine days) in the drinking water and feeding an alfalfa molt diet. Poultry Science 84: 1186-1190.

Moore, R.W., Park, S.Y., Kubena, L.F., Byrd, J.A., McReynolds, J.L., Burnham, M.R., Hume, M.E., Birkhold, S.G., Nisbet, D.G., Ricke, S.C., 2004. Comparison of zinc acetate and propionate addition on gastrointestinal tract fermentation and susceptibility of laying hens to *Salmonella enteritidis* during forced molt. Poultry Science 83: 1276-1286.

Onbaşilar, E.E., Erol, H., 2007. Effects of different forced molting methods on postmolt production, corticosterone level, and immune response to sheep red blood cells in laying hens. Journal of Applied Poultry Research 16: 529-536.

Park, S.Y., Woodward, C.L., Kubena, L.F., Nisbet, D.J., Birkhold, S.G., Ricke, S.C., 2008. Environmental dissemination of foodborne *Salmonella* in preharvest poultry production: Reservoirs, critical factors and research strategies. Critical Reviews in Environmental Science and Technology 38: 73-111.

Park, S.Y., Birkhold, S.G., Kubena, L.F., Nisbet, D.J., Ricke, S.C., 2004a. Effects of high zinc diets using zinc propionate on molt induction, organs, and postmolt egg production and quality in laying hens. Poultry Science 83: 24-33.

Park, S.Y., Kim, W.K., Birkhold, S.G., Kubena, L.F., Nisbet, D.J., Ricke, S.C., 2004b. Using a feed grade zinc propionate to achieve molt induction in laying hens and retain postmolt egg production and quality. Biological Trace Element Research 101: 165-179.

Park, S.Y., Kim, W.K., Birkhold, S.G., Kubena, L.F., Nisbet, D.J., Ricke, S.C., 2004c. Induced moulting issues and alternative dietary strategies for he egg industry in the United States. World's Poultry Science Journal, 60: 197-209.

Ricke, S.C. 2003. The gastrointestinal tract ecology of *Salmonella* Enteritidis colonization in molting hens. Poultry Science 82: 1003-1007.

Ricke S.C., Hume, M.E., Park, S.Y., Moore, R.W., Birkhold, S.G., Kubena, L.F., Nisbet, D.J., 2004a. Denaturing gradient gel electrophoresis (DGGE) as a rapid method for assessing gastrointestinal tract microflora responses in laying hens fed similar zinc molt induction diets. Journal of Rapid Methods and Automation in Microbiology 12: 69-81.

Ricke S.C., Park, S.Y., Moore, R.W., Kwon, Y.M., Woodward, C.L., Byrd, J.A., Nisbet, D.J., Kubena, L.F., 2004b. Feeding low calcium and zinc molt diets sustains gastrointestinal fermentation and limits *Salmonella enterica* serovar Enteritidis colonization in laying hens. Journal of Food Safety 24: 291-308.

Saengkerdsub, S., Anderson, R.C., Wilkinson, H.H., Kim, W.-K., Nisbet, D.J., Ricke, S.C., 2007a. Identification and quantification of methanogenic archaea in adult chicken ceca. Applied and Environmental Microbiology 73: 353-356.

Saengkerdsub, S., Herrera, P., Woodward, C.L., Anderson, R.C., Nisbet, D.J., Ricke, S.C., 2007b. Detection of methane and quantification of methanogenic archaea in faeces from young broiler chickens using real-time PCR. Letters in Applied Microbiology 45: 629-634.

Saengkerdsub, S., Kim, W.-K., Anderson, R.C., Woodward, C.L., Nisbet, D.J., Ricke., S.C., 2006. Effects of nitrocompounds and feedstuffs on in vitro methane production in chicken cecal contents and rumen fluid. Anaerobe 12: 85-92.

Schroeder, C.M., Latimer, H.K., Schlosser, W.D., Golden, N.J., Marks, H.M., Coleman, M.E., Hogue, A.T., Ebel, E.D., Quiring, N.M., Kadry, A.-R. M., Kause, J., 2006. Overview and summary of the food safety and inspection service risk assessment for *Salmonella* Enteritidis in shell eggs, October 2005. Foodborne Pathogens and Disease 3: 403-412.

Seo, K.H., Holt, P.S., Gast, R.K., 2001. Comparison of *Salmonella* Enteritidis infection in hens molted via long-term feed withdrawal versus full-fed wheat middling. Journal of Food Protection. 64: 1917-1921.

Soe, H.Y., Yayota, M., Ohtani, S., 2009. Effects of molt-induction period on induction of molt and post-molt performance in laying hens. Japanese Poultry Science 46: 203-211.

Soe, H.Y., Makino, Y., Uozumi, N., Yayota, M., Ohtani, S., 2007. Evaluation of non-feed removal induced molting in laying hens. The Journal of Poultry Science 44: 153-160.

Willis, W.L., Goktepe, I. Isikhuemhen, O.S., Reed, M. King, K., Murray, C., 2008. The effect of mushroom and pokeweed extract on *Salmonella*, egg production, and weight loss in molting hens. Poultry Science 87: 2451-2457.

Woodward, C.L., Kwon, Y.M., Kubena, L.F., Byrd, J.A., Moore, R.W., Nisbet, D.J., Ricke, S.C., 2005. Reduction of *Salmonella enterica* serovar Enteritidis colonization and invasion by an alfalfa diet during molt in leghorn hens. Poultry Science 84: 185-193.

Principles and application of glucogenic nutrient feed evaluation for ruminants

Wilfried M. van Straalen and Bart M. Tas
Schothorst Feed Research, P.O. Box 533, 8200 AM Lelystad, the Netherlands; wvstraalen@schothorst.nl; btas@schothorst.nl

Abstract

The first limiting nutrient for high yielding dairy cows is glucose. To predict the glucose supply and requirement in the cow a feed evaluation system was developed in which the glucogenic nutrients (GN) from propionic acid, digestible bypass starch and glucogenic amino acids can be calculated. Data for the fermentation and digestibility of the crude protein, starch, sugar and NDF in feedstuffs were based on nylon bag incubations in rumen and intestines. Tabulated values of GN for concentrate ingredients and calculation rules for roughages were developed to calculate GN supply in the ration. The rumen fermentation sub-model was validated using measurements on duodenal flow and volatile fatty acid proportion. The requirement for GN was based on the relationship between GN intake and response in milk yield and composition using feeding trial data. It was concluded that the GN feed evaluation system is a new tool for optimising diets for high yielding dairy cows towards increased milk, protein and fat production and was designed for practical use on farm level.

1. Introduction

For an optimal milk yield and composition dairy cow diets need to be balanced for first limiting nutrients. These nutrients are the product of the fermentation and digestion of feedstuffs in the digestive tract and metabolism in the animal. To obtain the right combination of feedstuffs in the diet, feed evaluation systems were developed. These systems describe the fermentation and digestion of the components of the diet and the sum of the digested nutrients can be compared with the requirement of the animal, mostly based on animal weight and milk production parameters.

Since some decades the energy supply and requirement of dairy cows is in most countries described in a Net Energy for lactation (NEl) system, like the VEM-system in The Netherlands (Van Es, 1978). Although this system has been quite successfully used to predict total milk production, it does not provide information on the composition of the milk. Because of the large difference between milk protein and fat price paid by the dairy industry and the milk fat quota system in the EU, optimising diets for milk protein and fat content has a large economic impact. This can not be achieved by the NEl systems. Another drawback of the NEl systems is that all feedstuffs are ranked according to the yield of net energy for milk production without discrimination of the type of energy. For the farmer it is very important whether the energy supplement in a diet comes from e.g. fermentable starch, that can result in rumen acidosis and subsequently in a drop in milk fat content, or from a bypass fat source, resulting in an increased fat content.

To overcome these drawbacks of the NEl system, the nutrient based feed evaluation system was developed at Schothorst Feed Research that predicts the availability of first limiting nutrients for high yielding dairy cows. These nutrients are described as glucogenic nutrients (for production of lactose), lipogenic nutrients (for production of milk fat) and aminogenic nutrients (for production of milk protein). The nutrient based feed evaluation system follows the fermentation and digestion pathways. In Figure 1 a simplified schematic presentation of the system is given. On regular diets about half of the milk fat production comes from long chain fatty acids that are digested in the small intestine or from mobilisation of body reserves. The other half comes from acetic acid and to a lesser extent from butyric acid that are the products of microbial fermentation of organic matter in the rumen and hind gut. Milk lactose is produced from glucose. Sources for glucose in the dairy cow are propionic acid from fermentation in rumen and hind gut,

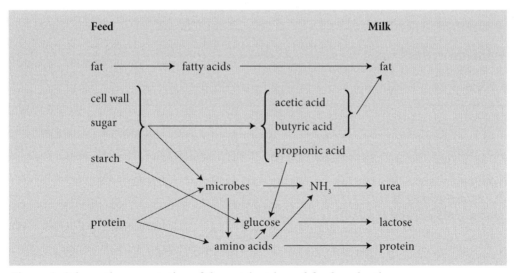

Figure 1. Schematic presentation of the nutrient based feed evaluation.

digested bypass starch in the small intestine and amino acids that are used for glucose production. Milk protein production is dependent on the availability of amino acids from microbial protein production in the rumen and from rumen bypass protein of feedstuffs.

2. First limiting nutrients

Glucose and methionine and lysine are regarded as the first limiting nutrients in high yielding dairy cows, because they are used as a substrate for the production of milk lactose and milk protein and as energy source for the production of milk fat and protein. In a meta analysis with literature data, Rulquin *et al.* (2007) showed that the protein to fat ratio increased with increased infusion of propionic acid in the rumen and with increased glucose infusion in the abomasum or small intestine (Figure 2). The nutrient based feed evaluation system was developed to enable optimisation of diets for these first limiting nutrients to predict more accurate milk production and milk composition. In this paper the background of the glucogenic nutrient approach and the application in practical diets is presented. The supply of first limiting amino acids is described as part of the true metabolisable protein system (TMP), that is not discussed in this paper. Although fatty acid composition of the feed can have a major impact on milk fat production and composition, the lipogenic nutrients are not regarded as first limiting for milk fat synthesis and are also not discussed in this paper.

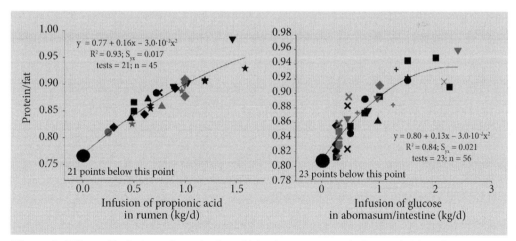

Figure 2. Effect of infusion of propionic acid in the rumen and glucose in the abomasum or small intestine on the milk protein to fat ratio (from Rulquin et al., *2007).*

3. Supply of glucogenic nutrients

As described above the glucogenic nutrients (GN) are the sum of glucose production from propionic acid from rumen and hindgut fermentation (GN-HPr), digestible bypass starch (GN-BST) and glucogenic amino acids not used for milk protein production (GN-AA). Because rumen fermentation of organic matter is responsible for the majority of the nutrient supply to the animal, the first step in the calculation of the glucogenic nutrient supply is the prediction of the fermentation and bypass of organic matter in the rumen and hindgut.

Fermentation of organic matter

The fermentation of carbohydrates and protein is predicted with the true fermentable organic matter (TFOM) system. This system is based on fermentation characteristics of feedstuffs measured with nylon bag incubation in the rumen of high yielding dairy cows and assumed passage rates for the different fractions of the organic matter. The carbohydrates are the sum of NDF, starch (ST), sugar (SU) and miscellaneous organic matter (REST = OM – CP – FAT – NDF – ST – SU – fermentation products). With nylon bag incubation the washable fraction (%W), undegradable fraction (%) and degradation rate (k_d) of the potential degradable fraction (%D) are determined for CP, starch, NDF and REST (for feedstuffs with REST > 100 g/kg). Also the water soluble fraction (%S) of protein is determined. The sugar fraction is assumed to be instantly degradable. Basal principles of the TFOM system were described by Van Straalen (1995).

An example for some concentrate feedstuffs is presented in Table 1. Feedstuffs with a slow protein fermentation are palmkernel meal and formaldehyde treated soybean meal. Starch in wheat shows a higher fermentation than in corn. Cell walls in palmkern meal have a relatively slow and incomplete fermentation. Also for roughages and wet by-products tabulated values of degradation parameters are available. For grass, grass silage and corn silage the value for the degradation parameters are dependent on the quality of the roughage determined by chemical analysis.

The assumed passage rate is dependent on the particle size of the feedstuff and component of the OM. For finely ground feedstuffs (2.5 mm) a passage of 6%/h and for roughages of 4.5%/h for all OM except NDF is assumed. The passage rate for NDF is assumed to be half of those of the other components of OM. For coarsely ground feedstuffs the passage rate is adapted towards the average particle size of the feedstuff.

Based on these degradation parameters and an assumed passage rate, for each feedstuff the true fermentable carbohydrates (TFCH) and

Table 1. Example of degradation parameters of CP, starch and NDF for feedstuffs.

	Crude protein				Starch		NDF	
	%S	%W	%U	k_d	%W	k_d	%U	k_d
Beetpulp	13	13	7	5.3	-	-	8	6.9
Citruspulp	33	40	4	5.0	-	-	10	6.2
Corn	5	18	2	3.0	22	5.3	3	1.6
Cornglutenfeed	10	51	5	6.0	32	11.5	6	3.4
Palmkernelmeal	4	16	8	2.5	-	-	24	5.3
Rapeseedmeal	5	15	7	10.8	-	-	32	9.3
Soybeanmeal	1	10	0	7.7	-	-	0	3.2
Soybeanmeal FT	2	14	1	1.3	-	-	0	3.2
Wheat	12	26	5	16.7	58	29.6	22	2.9

%S = soluble fraction; %W = washable fraction; %U = undegradable fraction; k_d = degradation rate.

protein (TFCP) is calculated. The true fermentable carbohydrates and protein are further subdivided in a rapidly fermentable part (RFCH and RFCP), moderately fermentable part (MFCH and MFCP) and slowly fermentable part (SFCH and SFCP). In Figure 3 the values for the different concentrate ingredients are presented. Due to the high washable fraction and degradation rate, wheat has a very high value for rapidly fermentable carbohydrates. Also beet pulp and citrus pulp have a high value for RFCH due to the high sugar content. Soybean meal and rapeseed meal have high values for fermentable CP.

Prediction of volatile fatty acid production

The second step in the nutrient supply is the prediction of the volatile fatty acid production in the rumen and large intestine. First the amount of substrate that is used for production of VFA is calculated by subtracting for every fraction of fermentable organic matter the amount used for microbial growth (Dijkstra, 1993). The amount of VFA produced from these substrates is calculated by using the coefficients determined by Bannink *et al.* (2006) (Table 2). Fermentation of starch results in a relatively high proportion of propionic acid while fermentation of sugar or NDF results in a higher proportion of acetic acid. Protein fermentation results in a higher proportion of other VFA (valeric acid, iso butyric acid and iso valeric acid). The total production of VFA is calculated by multiplication of the amount of each substrate with the proportions

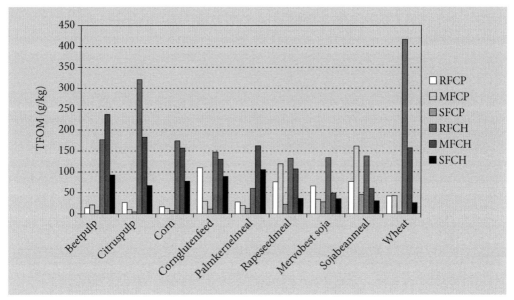

Figure 3. Example of fermentable carbohydrate fractions (RFCH, MFCH and SFCH) and protein fractions (RFCP, MFCP and SFCP) in concentrate ingredients.

of each VFA per substrate. It assumed that 35% of the propionic acid, 85% of the butyric acid and 15% of the acetic acid is metabolised by the rumen and intestinal wall (Bannink, 2000).

Digestible bypass starch

The amount of glucose from digested starch in the small intestine is calculated from the bypass starch content of the feedstuffs and the digestibility of bypass starch in the intestines. The bypass starch content is calculated in the TFOM system from degradation parameters and passage rates of starch for every feedstuff. The digestibility of

Table 2. Coefficients for calculation of VFA ratio per type of substrate.

Fermented substrate	Acetic	Propionic	Butyric	Other
Protein	51	26	11	12
Starch	49	26	19	7
Sugar/Rest	60	11	25	5
NDF	54	15	26	5

Adapted from Bannink *et al.* (2006)

bypass starch is assumed to decline with increased bypass starch level in the feedstuff (Nocek and Tamminga, 1991). For feedstuffs where this digestibility of bypass starch was measured with mobile nylon bag incubations, those values are used as tabulated values. An example of measured values is given in Figure 4.

Glucogenic amino acids

The amount of glucogenic amino acids that are not used for milk protein synthesis is calculated from the true metabolisable protein (TMP) not used for milk protein production (40%), corrected for the average ratio of glucogenic amino acids in protein.

Glucogenic nutrients

For the calculation of total glucogenic nutrients (GN) for each feedstuff the glucose supply from propionic acid from rumen and hind gut fermentation, digestible bypass starch and glucogenic amino acids using conversion factors for each nutrient towards glucose. In Figure 5 the GN for some concentrate ingredients are presented. The starch rich feedstuffs have a high level of GN, from fermentation of carbohydrates (wheat) or from digestible bypass starch (corn). For protein rich feedstuffs a large part of the GN comes from metabolisable amino acids not used for milk protein production.

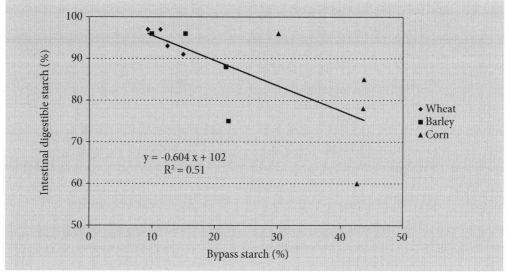

Figure 4. Relationship between bypass starch and intestinal digestion for wheat, barley and corn of different particle size.

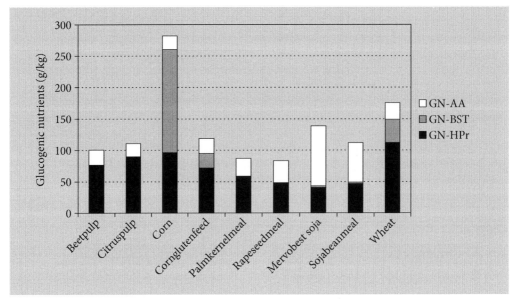

Figure 5. Glucogenic nutrients in concentrate ingredients estimated from propionic acid from rumen and hind gut fermentation (GN-HPr), digestible bypass starch (GN-BST) and glucogenic amino acids (GN-AA).

4. Validation of nutrient flow and VFA pattern

The nutrient based feed evaluation system predicts the flow of nutrients in the digestive tract. To validate the assumptions made to develop the model, a comparison was made between predicted nutrient flow at duodenal level and predicted VFA pattern in the rumen. For both studies datasets from literature publications were constructed. Publications were selected on availability of ration composition, feed intake and duodenal flow or rumen VFA parameters. The nutrient flow given in the publication was compared to the flow that was predicted by the nutrient based feed evaluation system.

The dataset to validate duodenal flow contained in total 140 treatments. A summary of intake and intestinal flow of different components of the DM is presented in Table 3. The number of data with intestinal flow varied between the different components.

Based on ration composition the flow of organic matter, protein, starch and NDF in the small intestine was calculated. The comparison between the determined and calculated values for OM, CP, starch and NDF (calculated with TFOM-system) is presented in Figure 6. The flow of organic matter is reasonable predicted with the TFOM system (R^2 = 0.80). With low contents of OM the predicted value corresponds better as compared with high levels, with high levels the model underestimates

Table 3. Overview of measured intake and intestinal flow of different components of DM in the dataset.

	DM	OM	CP	FAT	Starch	NDF
Intake (g/d)						
Number	140	133	136	50	96	116
Mean	18,588	16,919	3,190	836	4,925	6,327
Minimum	7,904	7,265	1,146	229	309	2,972
Maximum	25,500	23,400	5,333	1,550	9,231	8,700
Intestinal flow (g/d)						
Number	58	118	105	29	85	79
Mean	12,031	9,963	3,033	732	1,524	2,949
Minimum	5,163	3,890	1,332	311	121	468
Maximum	17,100	16,100	4,456	1,459	4,700	6,200

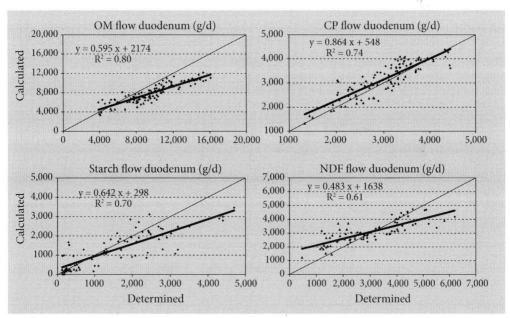

Figure 6. Comparison between determined (presented by the author) and calculated (based on ration) flow of OM, CP, Starch and NDF in the small intestine using the TFOM system.

the intestinal flow of OM. The regression line between predicted and determined CP flow was close to the line Y=X and showed a reasonable R^2 (0.74). The starch flow to the intestines is underestimated with the model, especially with higher levels of starch. The relationship between the predicted and measured starch flow was close (R^2=0.70). The NDF flow is overestimated at lower and underestimated at higher NDF levels. There is a reasonable relation (R^2 = 0.61).

The dataset to validate propionic acid ratio in rumen fluid contained data from 32 experiments and 106 treatments. Table 4 contains information of the dataset used for validation of the VFA pattern. In Figure 7 the relationship between predicted and observed propionic acid ratio in total VFA is presented. The observed values showed a larger variation than the predicted values and the propionic ratio was underestimated at the lower levels using the coefficients based on Bannink *et al.* (2006).

5. Determination of requirement for glucogenic nutrients

To determine the requirement for glucogenic nutrients for milk production and composition a dataset of feeding experiments was created. Data were used from 14 feeding experiments conducted at Schothorst Feed Research (SFR) from 1997 to 2004. Rations were calculated with current table values for nutritional characteristics. The dataset included both experiments in begin as in mid lactation. An overview of the measured production parameters and the calculated chemical composition and nutritional values are presented in Table 5. It was decided to use data from two periods with the same number of observations and with stable rations (week 7-9 and 20-22).

Table 4. Description of data set used for validation of VFA pattern.

	Mean	SD	Min.	Max.
DM intake (kg/d)	19.9	3.7	9.0	26.6
Milk (kg/d)	27.5	6.8	14.6	43.2
% roughage	52.6	14.6	10.0	99.9
Diet composition				
Crude protein	175	26	109	286
Fat	42	28	9	156
NDF	253	91	0	528
Sugar	42	15	22	96
Starch	339	69	206	651
VFA pattern (mmol/mol)				
Acetic acid	62.1	3.6	50.1	68.2
Propionic acid	22.2	3.8	15.9	37.0
Buteric acid	11.5	1.6	7.4	15.4
Other acids	4.2	1.4	1.3	8.0
pH	6.14	0.31	5.52	7.19

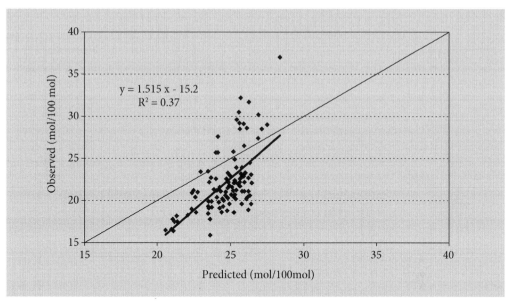

Figure 7. Relationship between measured and predicted propionic acid ration in total VFA.

Table 5. Mean, minimum and maximum values of production parameters and feed intake in the dataset.

n=2532	Mean	Min.	Max.	SD
Cows				
Parity	2.5	1	9	1.5
Stage of lactation (w)	14.5	7	22	6.6
Weight (kg)	627	457	885	69.7
BCS	3.1	2.0	4.0	0.3
DM-intake				
Roughage	14.2	6.9	23.3	2.2
Concentrates	9.3	4.0	12.7	1.8
Total	23.5	15.0	31.9	3.4
Production				
Milk (kg/d)	36.5	18.4	63.0	8.0
Fat (g/d)	1,532	781	2,669	310
Protein (g/d)	1,207	639	1,848	238
Lactose (g/d)	1,684	822	2,901	264
Urea (mg/dl)	24.0	8.4	38.9	5.0
Fat (%)	4.2	2.3	6.0	0.5
Protein (%)	3.3	2.7	4.3	0.2
Lactose (%)	4.6	4.1	5.1	0.2
FPCM (kg/d)	37.3	20.2	60.9	7.4

Multiple regression was used to determine the relation between production characteristics and nutritive value of the diets. This relation was calculated for the data of week 7-9 of lactation and week 20-22. In the model the stage of lactation (week), parity, weight and DM intake were used as covariates. The relations were fitted with a quadratic curve. The general model is:

$$Y = \beta_0 + lact + par + wght + DM + \beta_i \times FV_i + \beta_{ii} \times FV_i^2 + .. + + \beta_n \times FV_n + \beta_{nn} * FV_n^2$$

Where:
- Y = production characteristic; kg milk and fat, protein, lactose and urea (as % and g/d)
- β_0 = mean production characteristic
- lact = effect of stage of lactation (wk)
- par = effect of parity
- wght = effect of weight (kg)
- DM = effect of DM intake (kg/d)
- $FV_{i..n}$ = feeding value characteristic i to n
- $\beta_{i..n}$ = linear regression coefficient i to n
- $\beta_{ii..nn}$ = quadratic regression coefficient ii to nn

The regression coefficients obtained with this model were used to draw curves for the effect of glucogenic nutrients on animal performance (Figure 8). Effects of GN were more clear in early lactation than in mid lactation. Increase of GN nutrient in the diet resulted in an increased milk and milk protein production. The fat production showed an optimum at 135 g GN/kg DM diet. Based on these relationships a requirement of 125 g GN/kg DM diet for animals in begin lactation was established.

6. Practical application of the glucogenic nutrient system

The calculation rules of the glucogenic nutrient system were build into the SFR feed table that can be used for optimising the GN value for concentrates. For roughages the calculation rules are also implemented in ration calculation programs, so that the actual value for GN are calculated for each roughage available on farm level and diets can be optimised on GN. The GN level in the diet can be increased by enhancing rumen fermentation (GN from propionic acid) or to enhance the level of digestible bypass starch. Because increase in rumen fermentation is limited to avoid rumen acidosis, the most convenient way to increase GN is to enhance bypass starch.

As an example of the effects of the increase in GN from bypass starch in concentrate in two experiments carried out at SFR with animals in

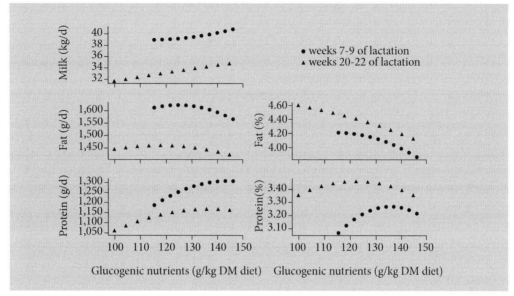

Figure 8. Relationship between glucogenic nutrients (in g/kg DM diet) and production parameters.

begin and mid lactation on feed intake and milk production parameters are presented in Table 6. In both experiments concentrates were fed according to a standard scheme while roughage was fed ad libitum. In begin lactation the basal diet consisted of 20% corn silage and 80% grass silage (in DM), while in mid lactation this ratio was 50/50. Within each experiment diets were equal for Net Energy lactation, True Metabolisable Protein and Rumen Protein Balance.

The increase in GN in the diet resulted in a higher milk, protein and lactose production while milk fat content decreased and milk fat production remained the same. The milk protein content increased during begin lactation and remained the same in mid lactation. The ratio between milk protein and milk fat was increased, resulting in a higher milk price.

7. Conclusions

The Glucogenic Nutrient feed evaluation system is a new tool for optimising diets towards increased milk and milk protein to fat ratio, resulting in a higher milk price. The system was designed for pratical use on farm level and follows the digestion pathways so that new knowledge can easily be incorporated.

Table 6. *Effect of GN on feed intake, milk production and composition in two experiments at begin and mid lactation (SFR, unpublished results).*

Lactation stage	Begin			Mid	
Treatment	1	2	3	1	2
Bypass starch in concentrate (g/kg)	5	30	45	5	35
Net Energy (MJ/kg DM diet)	6.9	6.9	6.9	6.7	6.7
Glucogenic nutrients (g/kg DM diet)	109	121	127	124	135
Intake (kg DM/d)	22.5	22.3	22.8	20.1	20.5
Milk (kg/d)	31.8	32.8	32.6	25.5	26.3
Protein (%)	3.36	3.44	3.45	3.50	3.51
Fat (%)	4.49	4.35	4.36	5.00	4.83
Lactose (%)	4.62	4.65	4.64	4.45	4.47
Protein (g/d)	1,063	1,122	1,122	891	917
Fat (g/d)	1,404	1,403	1,412	1,266	1,268
Lactose (g/d)	1,467	1,524	1,516	1,136	1,177

References

Bannink, A., Kogut, J., Dijkstra, J., France, J., Tamminga, S., Van Vuuren, A.M., 2000. Modelling production and portal appearance of volatile fatty acids in cows. In: McNamara, J.P., France, J., Beever, D.E. (eds.) Modelling nutrient utilisation in farm animals. CAB International, Wallingford, UK, pp. 87-102.

Bannink, A., Kogut, J., Dijkstra, J., France, J., Kebreab, E., Van Vuuren, A.M., Tamminga, S., 2006. Estimation of stoichiometry of volatile fatty acid production in the rumen of lactating cows. J. Theoret. Biol. 238: 36-51.

Dijkstra J., 1993. Mathematical modelling and integration of rumen fermentation processes. Thesis Landbouwuniversiteit Wageningen, 221 pp.

Nocek, J.E., Tamminga, S., 1991. Site of digestion of starch in the gastrointestinal tract of dairy cows and its effect on milk yield and composition. Journal of Dairy Science 74: 3598-3629.

Rulquin, H., Hurtaud, C., Lemosquet, S., Peyraud, J.-L., 2007. Effet des nutriments énergétiques sur la production et la teneur en matière grasse du lait de vache. INRA Production Animale 20: 163-176.

Van Es, A.J.H., 1978. Feed evaluation for ruminants. I. The systems in use from May 1977 onwards in the Netherlands. Livestock Production Science 5: 331-345.

Van Straalen, W.M., 1995. Modelling of N flow and excretion in dairy cows. Thesis Landbouwuniversiteit Wageningen, 205 pp.

Potential of fatty acids in mitigating ruminal methanogenesis

Veerle Fievez, Gunjan Goel, Charlotte Boeckaert and Bruno Vlaeminck
Laboratory for Animal Nutrition and Animal Product Quality, Department
of Animal Production, University of Ghent, Proefhoevestraat 10, 9090
Melle, Belgium; Veerle.Fievez@Ugent.be

Abstract

Dietary fat supplements rich in medium-chain fatty acid (MCFA) and poly-unsaturated fatty acid (PUFA) affect microbial activities, which can be exploited to mitigate ruminal methane emission through a direct effect on rumen methanogens or indirectly through defaunation or reduced substrate degradability. The anti-methanogenic activity of FAs is influenced by their dosage and the basal diet (forage:concentrate). However, the methane inhibitory effect of FA is often associated with negative effects on animal productivity by reduced feed intake and/or substrate degradability. This paper is the synthesis of various studies using methane inhibitory activity of FAs and discusses: a) sources and mode of action of dietary FA, b) methane inhibitory activity of MCFA and PUFA rich fat supplements from *in vitro* and *in vivo* studies, c) constraints in using these fat supplements for methane reduction at the expense of animal productivity and d) potential of these fat supplements to improve animal product quality.

1. Introduction

Microbial metabolism of plant carbohydrates and proteins in the rumen involves oxidation in an anaerobic environment with electron transfer to acceptors other than O_2. Electron sinks, involving pyridine nucleotide linked hydrogenases and H_2 production might provide maximal energy yield from the fermentation. However, under standard conditions, the regeneration of oxidised co-factors (NAD) from $NADH_2$ with the release of H_2 is thermodynamically unfavourable and only occurs when the concentration of H_2 is very low (Baldwin and Allison, 1983). Hydrogen transfer between fermenting and H_2 utilising bacteria prevents H_2 to accumulate (Wolin and Miller, 1988). In this respect, CH_4 production assures a very effective and continuous elimination of metabolic hydrogen, in the form of reduced protons, according to a reaction chain which can be summarised as: $CO_2 + 8H \rightarrow CH_4 + 2H_2O$ (Demeyer and

Van Nevel, 1975). Successive stages in this CO_2 reduction are mediated by unique enzymes and co-enzymes present in these methanogenic micro-organisms, the *Archaea* (Keltjens and Vogels, 1996). This removal of metabolic hydrogen ensures the regeneration of co-factors, which itself assures further oxidation of the substrate, stimulating degradation of cell wall carbohydrates and favouring growth of cellulolytic bacteria (Wolin and Miller, 1988; Fonty *et al.*, 1997). Co-factors are also regenerated and ATP is produced during the synthesis of the short-chain fatty acids (SCFA) propionate and butyrate, following the fermentation stoichiometry (Demeyer, 1991), with propionate being the predominant alternative hydrogen sink. In this context, organic acids (fumarate, malate, acrylate etc.) have shown to decrease methanogenesis (Carro and Ranilla, 2003) by acting as alternative H_2 acceptors and redirecting the flow of H_2 from methane formation through other reductive reactions such as those involved in propionic acid synthesis. Recently, Fievez *et al.* (2003) and Boeckaert *et al.* (2006) reported that PUFA rich fish oil and microalgae reduced methane by 80% and induced propionate production when studied *in vitro*. Increased proportions of propionate have also been observed in *in vivo* experiments using MCFA (e.g., Steele and Moore, 1968, Rohr *et al.*, 1978 and Machmüller *et al.*, 2003a) indicating that at least part of the available hydrogen was incorporated into propionate. Therefore, dietary fat supplements rich in MCFA and PUFA could also serve as strategies to mitigate ruminal methane production due to toxic effects of free fatty acids (FFA) on both methanogens (Prins *et al.*, 1972) and/or associated protozoa (Czerkawski *et al.*, 1975). However, bactericidal properties of FAs have been suggested by Nieman (1954), therefore, consideration should be given to the fat content in the ruminant diet to avoid negative effects on animal productivity (Jenkins and Palmquist, 1984). Nevertheless, conscious application of lipid supplements might be an effective and realistic strategy to specifically target rumen methanogenesis, although Eugène *et al.* (2008) concluded methane reduction to be mainly linked to a decreased DMI with lipid supplemented diets. Nevertheless, no distinction between types of fat has been made in this meta-analysis study. In the current paper, we will discuss the potential of specific FA sources, particularly those rich in MCFA or PUFA to reduce rumen methanogenesis.

2. Chemical structure of MCFA and PUFA and their sources

The FA can be categorized into three types based on their structure and chain length, short-chain fatty acids (SCFA) consisting of C2-C6, medium-chain fatty acids (MCFA) C8-C14, long-chain fatty acids (LCFA) C16-C24, including poly-unsaturated fatty acids (PUFA) of chain length of 18 or more carbon atoms with higher degree of double bonds. Their prevalence in natural oils and oil sources, with emphasis on particular sources discussed in the current text, is mentioned in Table 1.

3. General antibacterial action of MCFA and PUFA

Nieman (1954) concluded that, FAs with a chain length of around C12 were most active as antibacterial agents. Further, the inhibitory activity of unsaturated FAs was more pronounced than that of saturated acids, with gram positive bacteria being more susceptible to the action of FAs than the gram negative bacteria. This is linked to differences in the cell wall structure of both types of bacteria, which is more impermeable for gram negative bacteria. Various factors influencing the antimicrobial activity of FAs include, (1) pH of the medium, the activity being higher in an acid environment, (2) presence of calcium ions (Galbraith *et al.*, 1971), (3) chain length (Jenkins and Palmquist, 1982), hydrophobicity of the hydrocarbon chain and the presence of a free carboxyl group as well as iv) presence of particulate phase, determining the adsorption rate of the FA onto either the particulate material or the bacteria, which inhibits bacterial metabolism.

The saturated MCFA and PUFA differ in their mechanism of action. The MCFA diffuse into bacterial cells in their undissociated form and dissociate within the protoplasm, thereby leading to intracellular acidification. A lower intracellular pH can lead to inactivation of intracellular enzymes and inhibition of amino acid transport and uncouple ATP-driven pumps, leading to cell death (Freese *et al.*, 1973). At sublethal concentrations, these can interfere with bacterial signal transduction, and inhibit the expression of virulence factors and antibiotic resistance genes (Nair *et al.*, 2005).

Galbraith *et al.* (1971) related the antibacterial activity of C18 fatty acids to unsaturation and isomerism as follows: C18:0 < C18:1 *trans* < C18:1 *cis* < C18:2 ≤ C18:3. They suggested that the difference in the order of activity of these C18 fatty acids correlates with differences in solubility and interfacial behaviour. The toxic effect was suggested to be due to the adsorption of surface active lipids on the microbial cell wall (Czerkawski *et al.*, 1966a). Saturated and unsaturated FA seem to exert different properties in this respect as demonstrated in anaerobic sludge (Pereira *et al.*, 2005). Methanogenic activity was detected in sludge to which C16:0 was added, which precipitated in white spots. However, methane production of sludge to which C18:1 cis-9 was added only occurred 50h after being washed with ethanol. This suggests a reversible encapsulation of the microbes by unsaturated fatty acids. However, besides (reversible) physical adsorption, some (unsaturated) fatty acids readily incorporate into membrane phospholipids of the bacterial cells, as shown for DHA (Stillwell and Wassall, 2003). Once esterified into

Table 1. Fatty acid composition of selected oils used in animal feeds, with some particular emphasis on composition of sources used in the current paper (linked to data of Figure 2[1]) (Fievez et al., 2003; Boeckaert et al., 2007, Panyakaew et al., 2008; Anonymous, 2009).

Oil Source	% of oil in seed/ kernel	Caproic C6:0	Caprylic C8:0	Capric C10:0	Lauric C12:0	Myristic C14:0
Coconut oil	63-65	0-0.8	5.0-9.0	6.0-10.0	44.0-52.0	13.0-19.0
Krabok oil	----	----	----	1.6	42.0	46.4
Palm Kernel oil	44-65	tr	3.0-5.0	3.0-7.0	40.0-52.0	14.0-18.0
Palm oil	30-60	----	----	----	----	0.5-2.0
Rapeseed oil[2]	33-40	----	----	----	----	----
Rice Bran oil	15-23	----	----	----	----	0.4-1.0
Corn (maize) oil	48	----	4	7	----	0.2-1.0
Soybean oil[1]	50	----	----	----	----	0.1 (0.1)
Sunflower oil	25-35	----	----	----	----	----
Safflower oil	25-37	----	----	----	----	tr.0.5
Linseed oil[3]	40-44	----	----	----	----	----
Fish oil[1]	----					7.7-8.4 (7.7)
Micro algae[1]	42-57					9.3-11.8 (9.3)

[1] Composition of oil sources used for Figure 2 data is given between brackets. The oil fraction of sunflower seed and linseed were assumed to have a similar composition as their respective oils for data calculation in Figure 2.

phospholipids, many basic properties of membranes, including acyl chain order and fluidity, phase behaviour, elastic compressibility, permeability, fusion and protein activity might be significantly altered (Stillwell & Wassall, 2003). This is suggested to be due to an altered shape of the unsaturated fatty acid molecule (kinked) which disrupt the lipid bilayer structure of microorganisms (Keweloh and Heipeiper 1996). Differences in bacterial cell wall structure might explain higher or lower sensitivity of specific bacterial groups (Prins *et al.*, 1972; Maczulac *et al.*, 1981).

Other (additional) working mechanisms, specifically linked to rumen methane inhibition also have been suggested and will be discussed in the next section.

Palmitic C16:0	Stearic C18:0	Oleic C18:1	Linoleic C18:2n-6	Linolenic C18:3n-3	EPA C20:5n-3	DHA C22:6n-3
8.0-11.0	1.0-3.0	5.0-8.0	0-1.0	----	----	----
4.5	0.4	2.6	0.3	0.04	----	----
7.0-9.0	1.0-3.0	11.0-19.0	0.5-2.0	----	----	----
32.0-45.0	2.0-7.0	38.0-52.0	5.0-11.0	----	----	----
1.5	0.4	22	14.2	6.8	----	----
12.0-18.0	1.0-3.0	40.0-50.0	29.0-42.0	0.5-1.0	----	----
8.0-12.0	2.0-5.0	19.0-49.0	34.0-62.0	----	----	----
3.1-11.6	3.7-3.8	24.0-32.7	53.2-54.2	1.13-6.2	----	----
(11.6)	(3.7)	(24.0)	(54.2)	(6.2)		
3.0-6.0	1.0-3.0	14.0-35.0	44.0-75.0	----	----	----
3.0-6.0	1.0-4.0	13.0-21.0	73.0-79.0	tr.	----	----
4.0-7.0	2.0-5.0	12.0-34.0	17.0-24.0	35.0-60.0	----	----
13.5-17.0	1.2-3.4	12.5-15.0	1.8-3.6	1.0-1.2	5.8-18.7	7.6-11.7
(17.0)	(3.4)	(15.0)	(3.6)	(1.0)	(18.7)	(11.7)
16-9-20.7	1.92-8.42	15.9-16.7	0.8-2.5	0.07-0.5	0.04-0.05	35.1-49.3
(16.9)	(1.92)	(16.7)	(2.5)	(0.07)	(0.05)	(49.3)

[2] Alternative name: mustard seed oil.
[3] Alternative name: flax seed oil.

4. Methane inhibitory mechanism of MCFA and PUFA

Defaunating effect of MCFA and PUFA

Both MCFA and PUFA rich lipids have been suggested as defaunating agents (Newbold and Chamberlain, 1988), indirectly resulting in reduction of rumen methanogenesis. This is determined by several factors including the loss of methanogens attached to the protozoa, which are responsible for 9 to 25% of the CH_4 production in the rumen (Newbold et al., 1995). Additionally, defaunation will reduce the production of CH_4 precursors both through the loss of protozoa which produce important quantities of H_2 and formate (Williams and Coleman, 1992) and by a lower rumen digestibility of the crude fibre fraction (De

Smet, 1993). However, the effects of MCFA and PUFA on CH$_4$ production are certainly not limited (Van Nevel and Demeyer, 1995) and sometimes not even induced by those mediated *via* the rumen protozoa. Indeed, Machmüller *et al.* (2003b) showed reduction of methane by coconut oil in both faunated (with ciliate protozoa) and chemically defaunated (without ciliate protozoa) rumen fluid which was later on confirmed under *in vivo* conditions also. Similarly, Fievez *et al.* (2003) observed a reduction in protozoal counts in only 1 of 4 fish oil supplemented sheep, despite a similar rumen CH$_4$ reduction among the sheep.

Methane inhibitory through reduction of intake and/or rumen degradability

Reduction of methane by MCFA or PUFA could be related to a reduction in feed intake (e.g. Eugène *et al.*, 2008) and/or fermentation (e.g. Demeyer and Fievez, 2000). Obviously, this methane inhibitory mechanism is not desirable as this means a reduced productivity. Although the absolute amount of methane excreted is relevant, e.g. in terms of total greenhouse gas emission from livestock, the amount of methane produced per unit of substrate degraded represents the true efficacy of a manipulation strategy. Therefore, in the next parts, methane reduction will be considered per unit of apparently rumen degraded organic matter (mainly *in vitro* studies as these data were scarce under *in vivo* conditions) or per unit of DMI (*in vivo*).

Direct methane inhibitory effect of MCFA

Direct toxicity of MCFA against methanogens have been reported by Soliva *et al.* (2003) *in vitro* using H$_2$ and CO$_2$ as substrate for methane production (to avoid undesired interactions of MCFA with feed particles). They reported a 78% and 31% decrease in ruminal archaea with C12:0 and C14:0 respectively. Furthermore, the supplementation of C12:0 and C14:0 affect the methanogenic diversity to different extents. Methanococcales, the most abundant methanogenic order in the former study, declined to a larger extent than the other orders. Supplementation with C12:0 alone, decreased the counts of Methanococcales, Methanomicrobiales, Methanosarcinales and Methanobacteriales by 74, 73, 66 and 55%, respectively, compared with the unsupplemented control.

Direct methane inhibitory effect of PUFA

The reduction of rumen methanogenesis seems to respect the general 'antibacterial' order of unsaturated fatty acids, with lower minimum methane inhibitory concentrations for C18:3, followed by C18:2 and C18:1 (Czerkawski *et al.*, 1966b). Methane inhibition by PUFA is further determined by a free carboxyl group, explaining why the rumen methane reduction of a broad variety of triacylglycerols seems to be dependent on their rate of lipolysis (Van Nevel, 1991; Fievez *et al.*, 2003). As lipolysis also represents a prerequisite of PUFA biohydrogenation, direct competition for H_2 between methanogens and hydrogenating bacteria is often thought to be the driving force for methane suppression by PUFA. However, unsaturated fatty acids inhibit rumen methanogenesis at millimolar concentration (Prins *et al.*, 1972), which is far too low to serve as major competitors for hydrogen. In this respect, Fievez *et al.* (2003) calculated that less than 1% of the hydrogen, released after methane inhibition by fish oil, was used for the biohydrogenation of PUFA. Adversely, incubation under a H_2-rich atmosphere did not eliminate the inhibitory effect of DHA (Fievez *et al.*, 2007).

Further bactericidal effects of PUFA have been attributed to free radicals, originating from (auto) oxidation of PUFA (Knapp and Melly, 1986). This might be of particular importance in an anaerobic environment such as the rumen, where only facultative anaerobes express superoxide dismutase activity (Holovska *et al.*, 2002). Strict anaerobes, such as the methanogens, lack antioxidative enzymes and hence bacterial defence mechanisms against oxidation products (McCord *et al.*, 1971). If PUFA would indirectly inhibit rumen methanogenesis through oxidation, addition of external antioxidants might be hypothesised to alleviate this, similarly to their effect on rumen microbial growth and fibre digestion (Hino *et al.*, 1993). Although this needs further experimental evidence, preliminary experiments showed rumen methane production during in vitro incubations with (oxidized) DHA did not recover through supplementation of vitamin C or E (Boeckaert *et al.*, 2007).

5. Quantitative effect of MCFA and PUFA or their oil sources on rumen methane production

Effects of MCFA or their sources

Dohme *et al.* (2001) compared the efficacy of individual fatty acids (C8-C18) *in vitro* using the Rusitec system and concluded that C12:0 and C14:0 are the only MCFA effective in suppressing rumen methanogenesis and methanogenic counts. Nevertheless, in a recent study of our group, C10:0 has been shown to effectively suppress rumen methanogenesis (Goel *et al.*, 2009). Still, most of the studies with MCFA focussed on sources particularly rich in C12:0 and C14:0. Soliva *et al.* (2003) reported direct effects of MCFA on rumen archaea, where both methane production and numbers of rumen methanogens declined curvilinearly with increasing proportions of non esterified C12:0 to C14:0 in the FA mixtures using H_2 and CO_2 as substrate (4:1). A proportion of

2:1 (C12:0/C14:0) possessed the same maximum methane-suppressing effect (96%) as with C12:0 alone indicating the synergistic effect of C14:0 to enhance the methane-suppressing effect of C12:0 in mixtures having a C12:0/C14:0 ratio between 14:1 and 2:1. In contrast, a batch *in vitro* study by Panyakaew *et al.* (2008) showed similar anti-methanogenic effect of coconut and krabok oil, supplemented to reach equal amounts of C12:0 + C14:0. In the krabok oil supplemented incubation, this means a relatively higher supply of C14:0 due to the higher proportions of this FA in the krabok oil compared to the coconut oil (20.5 vs. 46.4 g/100 g total FA for coconut and krabok oil, respectively). The results of this study suggest that both C12:0 and C14:0 have a similar inhibitory activity on methanogenesis, opposite to the results of Soliva *et al.* (2004). Future investigations should clarify whether and why synergistic effects between specific mixtures of non-esterified C12:0 and C14:0 occur in the inhibition of ruminal methanogenesis.

In vivo studies have been performed both using pure MCFA as well as MCFA rich sources. Machmüller *et al.* (2006) reported that a 50% reduction in *in vivo* methane could be possible at a dietary inclusion of 3% of MCFA of C12:0 and C14:0. Supplementation of MCFA (C12:0 and C14:0) reduced the methane emission by 0.16 ± 0.03 g per day per $kg^{0.75}$ live weight for each g of MCFA supplied per kg of dietary DM. In terms of gross energy intake, the unsupplemented diets, results in emission of on average $6.32 \pm 0.25\%$ of the gross energy intake as methane. For each g of MCFA C12:0 and C14:0 supplied (g per day per $kg^{0.75}$ live weight), the energy loss via methane (percentage of gross energy intake) was reduced by 2.13 ± 0.22.

Further, inclusion of MCFA sources showed considerable variation in methane as illustrated in Figure 1, representing the effects of different MCFA sources studied *in vivo* in sheep and heifers., Overall, an average methane reduction of 5% per % of dietary inclusion of MCFA sources was observed. However, in some studies (e.g. Machmüller and Kreuzer, 1999, not included in Figure 1) oil inclusion level and dietary concentrate proportion were confounded. Indeed, the 63.8% methane reduction at a coconut oil inclusion rate of 7% (on DM basis) should be partially attributed to a higher dietary proportion of concentrate in the latter study as a lower methane emission rate is associated with concentrate as compared to roughage. Additionally, from the compilation of literature data (Figure 1) a synergistic effect between MCFA and dietary concentrate proportions could be suggested. Indeed, MCFA seem to lose their anti-methanogenic activity when included in a basal diet consisting of more than 55% of forages. Inversely, supplementation of MCFA sources to reduce rumen methanogenesis might be suggested to be more effective in concentrate based diets.

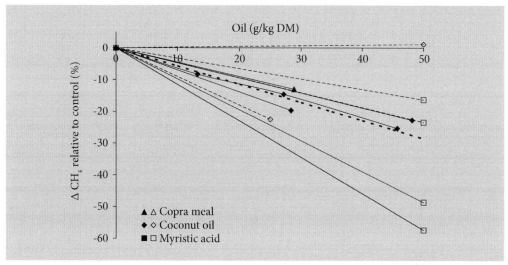

Figure 1. Effect of different levels of MCFA rich sources (g/kg DM) on relative methane reduction with different basal diets (the dashed lines indicate the studies with basal diet with more than 55% forages, the solid lines indicate basal diet with more than 45% concentrate; sheep: open symbols; beef heifer: closed symbols. The thick dashed line represents the average effect. Data from Machmüller et al. *(2003a); Lovett* et al. *(2003); Machmüller* et al. *(2003b); Jordan* et al. *(2006a); Machmüller* et al. *(2001); Jordan* et al. *(2006b).*

Effect of PUFA or their oil sources

The effect of unsaturated FA on methane production seems directly proportional to the degree of unsaturation as suggested from older *in vitro* studies using unsaturated C18 FA (Czerkawski *et al.*, 1966b; Demeyer and Henderickx, 1967) (Figure 2). More recent studies with pure, unesterified oleic, linoleic and linolenic acid confirmed this relation, but the C18 PUFA showed a numerical stronger effect, whereas the inhibitory effect of oleic acid was weaker (Zhang *et al.*, 2008; data not shown) as compared to the study of Czerkawski *et al.* (1966b). From a practical point of view, the methane inhibitory effect of C18:2 and C18:3 rich sources (oils or seeds) is of higher relevance, but highly variable rates of inhibition were observed *in vitro* (Figure 2). Marine oil sources (fish oil and micro algae) seem the most effective methane inhibitors (Figure 2), in line with their higher degree of unsaturation (Table 1). Although direct extrapolation of methane production (l/kg DMI) from *in vitro* to *in vivo* should not be emphasized (Flachowsky & Lebzien, 2009). A literature survey of *in vivo* experiments, involving thirty-seven diets (seven papers), confirmed the relation between CH_4 diminution and degree of unsaturation of C18 fatty acids in the dietary fat (Giger-Reverdin *et al.*, 2003).

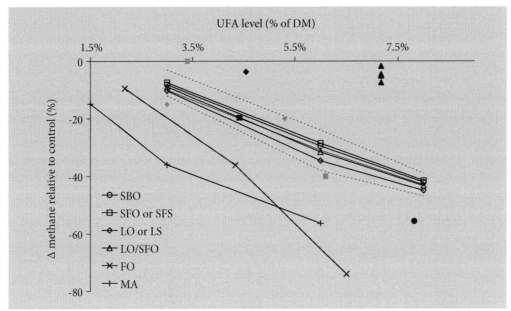

Figure 2. In vitro *methane inhibition (expressed relative to iso-lipidic control) related to level of unsaturated fatty acids (% of DM) Dotted lines are data from Czerkawski et al. (1966b) using pure unesterified C18:1 n-9 (upper), C18:2 n-6 (middle) and C18:3 n-3 (lowest). Full lines with open symbols represent potential methane inhibition by oil sources as assessed from extrapolations of Czerkawski's results based on their fatty acid composition (Table 1). Full symbols represent effectively measured methane inhibition, with grey symbols representing sunflower (SFS) or linseed (LS) and black symbols indicating soybean oil (SBO), sunflower oil (SFO) or linseed oil (LO) (data from Machmüller et al., 1998; Fievez et al., 2003; Fievez et al., 2007 and Hassim et al., 2008). Fish oil (FO) and micro algae (MA) data also represent effectively measured inhibition (Fievez et al., 2003; Fievez et al., 2007).*

In accordance with the *in vitro* studies, FA with 20 or more carbon atoms showed a dramatic impact on CH_4 emissions, but no specification on the type of fatty acids within the latter broad term were included as this information was lacking in the feed tables on which the authors relied (Giger-Reverdin *et al.* 2003). Methane emission was assessed as:

$$CH_4 \ (l/kgDMI) = 46.3 - 0.0188DMI^2 - 0.545 \ index - \\ - 91.1 \ \Sigma C_{\geq 20} \ (R^2 = 0.84, n = 37, R.S.D. = 2.16 \ l/kg \ DMI),$$

with index of unsaturation being the quantity of each of the unsaturated FA multiplied by its degree of unsaturation and the $\Sigma C_{\geq 20}$ term corresponding to the sum of FA with 20 to 26 C atoms. From the few direct fish oil supplemented *in vivo* measurements (n=3), it is suggested that the equation by Giger-Reverdin *et al.*(2003) overestimates the methane inhibitory potential of those products, at least at dietary inclusion levels of 30% and beyond (Figure 3). The deviation of the actual data of Woodward *et al.*(2006) from the predicted inhibition (Giger-Reverdin *et al.*, 2003) also might be due to the use of fish oil with a FA composition different from the one mentioned in Table 1. However, no information on the fish oil fatty acid composition was given by Woodward *et al.* (2006). The actual data of our study (Fievez *et al.*, 2003) relied on indirect methane assessments with sheep, which might be another reason for deviation from the predicted inhibition (Giger-Reverdin *et al.*, 2003). Obviously, more data are required with marine products. Further, in some cases registered methane inhibition through supplementation of oilseeds to dairy cows deviated from the expected values (Figure 3), which might be linked to differences in technological treatment of the oilseeds. E.g. in the studies presented in Figure 3, this included both untreated, crushed and extruded linseed. It also should be noted that dietary inclusion levels of the studies with direct measurements (Martin *et al.*, 2008; Beachemin *et al.*, 2009) are considerably higher than applied in practice. Extrapolation, based on the relation of Giger-Reverdin *et al.* (2003), of the methane reduction induced by daily supplementation of 3.5 kg extruded linseed (-26%; Martin *et al.*, 2008) to a more practical amount of 1.5 kg extruded linseed, revealed a potential reduction of about 18%. Obviously, this should be confirmed by direct measurements.

6. Interactions of fat sources with the basal diet

A positive response on methane reduction to high levels of starch-based concentrate (grains) has also been reported by others (Beauchemin and McGinn, 2005). An increased proportion of starch in the diet changes ruminal volatile fatty acid (VFAs) concentrations in such a way that less acetate and more propionate is formed, and the supply of hydrogen for methanogenesis is limited. Also, the pH decreases as the proportion of propionate increases, which reduces methanogenic activity (Walichnowski and Lawrence, 1982). Additionally, concentrate feeding has shown to reduce methane output by reducing the protozoal population (Van Soest, 1982). Moreover, in several cases an interactive effect between methane reducing capacity of an oil supplement and the basal diet has been suggested. E.g. both *in vitro* (Machmüller *et al.*, 2001, 2002) and *in vivo* (Machmüller *et al.*, 2003a, 2003b) interactions with the dietary content of minerals (i.e. Ca) and fibre (i.e. composition of the carbohydrate fraction) influence the methane-suppressing effect of oils. Machmüller *et al.* (2001), in their *in vitro* Rusitec study, reported higher effect of MCFAs from coconut oil on methanogenesis (reduction by 62%) with an intensive type of basal diet (with low structural carbohydrate content) as compared to an extensive type containing high structural

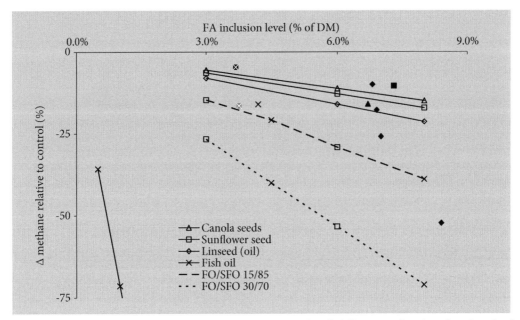

Figure 3. Methane inhibition (expressed relative to iso-lipidic control) related to level of supplemented fat (% of DM). Lines with open symbols represent potential methane inhibition by oilseeds rich in unsaturated C18 fatty acids, assessed as from CH_4 inhibition (%) = -0.545 * index/ 38.8 according to Giger-Reverdin et al. (2003). Full symbols represent effectively measured methane inhibition upon supplementation of canola seeds, sunflower seeds or linseed, either untreated, crushed or extruded (data from Beauchemin et al., 2009; Martin et al., 2008). Similarly, connected crosses represent potential (Giger-Reverdin et al., 2003) methane reduction by fish oil (CH_4 inhibition (%) = -91.1 × $\Sigma C_{\leq 20}$/ 38.8), whereas the individual symbol represents the effectively assessed methane reduction in sheep (Fievez et al., 2003). Lines with squares and crosses represent fish oil/ sunflower oil (FO/ SFO) combinations according to Giger-Reverdin et al. (2003). Full squares with crosses are actual measurements (grey, 15/ 85 and black, 30/ 70) (Woodward et al., 2006).

carbohydrate content (6% reduction). Similarly, contrasting effects of capric acid were noticed due to different basal diets from two *in vitro* studies: Goel *et al.* (2009) and Dohme *et al.* (2001), where the former authors reported methane inhibitory effects of capric acid in concentrate based diets whereas no effect of capric acid on methane production was observed when supplemented to a basal diet consisting of 60% forage (Dohme *et al.*, 2001) or 100% hay in our study (no detailed data shown by Goel *et al.*, 2009). Recently, Klevenhusen *et al.* (2009) also reported a diet-dependent effect of monolaurin, which was effective in reducing methane production with straw-grain diets whereas no effect was observed with the hay based diet. These studies confirmed that inclusion of roughage in basal substrate reduces the methane suppressing effect of MCFAs. To our knowledge, this interaction has not been studied with PUFA sources. Nevertheless, similar results might be expected as more extensive biohydrogenation (and hence, disappearance) of the toxic PUFA has been reported when supplemented to high fibre diets (Dang Van *et al.*, 2008).

Further, *in vitro* and *in vivo* studies from Machmüller *et al.* (2001, 2002) highlighted the importance of dietary Ca levels as they obtained greater reductions with C14:0 when dietary Ca levels were according to the actual requirements of the animal. High levels of Ca in the diets counterbalanced the action of MCFAs. Therefore, caution should be taken for Ca levels in formulating a feed supplemented with MCFA or PUFA as feeding strategies against methane formation. Generally, these interactions could result in a decline of freely available fatty acids in the (simulated) rumen environment, which has been indicated as the crucial factor determining the inhibitory effect of MCFA on pure cultures of *Methanobrevibacter ruminantium* (Henderson, 1973).

Furthermore, addition of MCFA in unesterified form results in more reduction of methane as compared to addition as esterified forms. Indeed, in the latter case, the rate of lipolysis then determines the effectiveness of a particular MCFA in reducing methane. Differences in rate of lipolysis also might be linked to variation in methane inhibition of various technologically treated oilseeds (cfr. Figure 3) (e.g. Doreau *et al.*, 2009).

7. Side effects of oil sources

Feed Intake and degradability

Inclusion of fat in animal diets increases the energy density of the diet but this does not always lead to improved animal performance (Lovett *et al.*, 2003; Jordan *et al.*, 2006a). The decrease in fibre degradability was reported for MCFA *in vitro* Dohme *et al.* (2001) and *in vivo* (Ushida *et al.*, 1992). Dohme *et al.* (2001) reported 35% reduction in apparent degradation of NDF using C12:0 (50g/kg DM). The decrease in fermented organic matter by approximately 25% and 50% was also noticed *in vitro* (Zhang *et al.*, 2008) using C18:3, added at a level of 35 and 70 g/kg DM. Also, *in vitro* studies with PUFA sources generally show a reduction in apparent OM fermentation at higher PUFA inclusion levels (Fievez *et al.*, 2003; Fievez *et al.*, 2007).

In vivo, the soapy taste of MCFA might reduce intake. However, the effect of MCFA supplementation on DMI is absent or limited when the basal diet included more than 55% of forages compared to the studies with more than 45% concentrate (Figure 4). As methane reduction was limited when no effect on DMI was observed (Figure 1), effects on DMI most

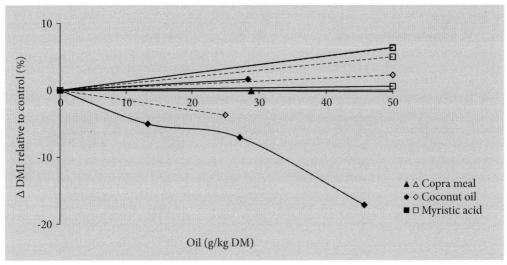

Figure 4. Effect of different levels of MCFA rich sources (g/kg DM) on relative DMI (g/kg) with different basal diets (the dashed lines indicate the studies with basal diets with more than 55% forages, the solid lines indicate basal diets with more than 45% concentrate; sheep: open symbols, beef heifer: closed symbols. Data from Machmüller et al. (2003a); Lovett et al. (2003); Machmüller et al. (2003b); Jordan et al. (2006a); Machmüller et al. (2001); Jordan et al. (2006b).

probably are induced through adverse effects on rumen fermentation rather than intrinsic characteristics of the oily supplements. *In vivo* studies using PUFA rich oil seeds from Beauchemein *et al.* (2009) and Martin *et al.* (2008) also reported a decrease in DMI and organic matter digestibility with supplementation of different forms of linseed (Table 2). It should be noted, that, although pure oils are more effective in inhibiting methane than the same amount of fat supplied through different forms of seeds (Figure 3), seeds might be preferred because of their lower risk to induce side-effects on feed intake and fibre digestibility.

Effect on production parameters, with emphasis on milk fatty acid composition

Dietary lipids generally increase milk yield as reviewed by Chilliard and Ferlay (2004). Dohme *et al.* (2004) and Odongo *et al.* (2007) did not report any affect of myristic acid (C14:0) on milk yield at 4% and 5% levels, respectively. However, when DMI and/or digestibility is suppressed (e.g. Table 2), milk yield is reduced. Further, PUFA also have been suggested to reduce milk fat content and/or yield (e.g. Table 2). On the other hand, these FA sources recently received much attention with respect to the incorporation of beneficial monounsaturated FA (MUFA), PUFA and conjugated linoleic isomers in milk fat (Figure 5a). With unprotected

Table 2. Effect (relative to control) of dietary supplementation of linseed (oil) (dietary supplementation rate of 7.4 to 8.4% of DM) on DMI, animal performance and digestibility.

Δ (%)	Crushed [1]	Crushed [2]	Extruded [2]	Oil [2]
DMI	+1.6	-1.5	-15.6	-25.8
Milk yield	-2.9	-1.3	-19.2	-27.8
Milk fat content	-0.62	+10.5	-14.1	-	-21.4
OM digestibility	-8.50	-6.86	-4.71	-6.57

Source: [1] Beauchemin et al. (2009); [2] Martin et al. (2008)

C18-PUFA sources (e.g. linseed, sunflowerseed, soybean), increases in milk PUFA content are modest and benefits are mainly situated in increases of MUFA (oleic acid). Similar shifts might be obtained with – less expensive – dietary MUFA sources, e.g. rapeseed (Figure 5a). However, effects on rumen methanogenesis of these less unsaturated fatty acid sources are expected to be lower (cfr. Section 4). Similarly, a higher impact on milk PUFA content (with the sum of C18:2n-6 and C18:3n-3 representing up to 10% of the milk fat) might be reached when using protected fats (Figure 5b). Moreover, protected fats will alleviate the negative side effects on rumen metabolism. However, this protection will limit the effect on the methane producing microorganisms and hence their potential as methane mitigation strategies. Additionally, overprotection of these fat supplements due to different chemical and physical treatments should be considered before their use in ruminant diets. With regards to marine PUFA sources (e.g. fish oil or marine algae), their potential for methane reduction seems high, but incorporation of these long PUFA (mainly EPA and DHA) in milk fat is minor (Figure 5c), both due to rumen biohydrogenation as well as predominant transport in the blood plasma phospholipid fraction, for which the udder lipoprotein lipase only shows a low affinity.

From their LDL cholesterol increasing capacity, lauric (C12:0), myristic (C14:0), palmitic and trans fatty acids were considered the most noxious of fats. In this respect, supplementation of MCFA sources, particularly rich in C12:0 and C14:0, might be undesirable as they are transferred to the milk fat (Figure 5d), whereas PUFA sources increase the trans fatty acid content (Figure 5a-c). However, as fats also affect HDL cholesterol, which seem to lower the risk for cardiovascular diseases, the ratio total to HDL cholesterol has been suggested more recently as a better marker (Mensink et al., 2003). Based on the latter marker, trans fatty acids were classified twice as detrimental compared with saturated fatty acids. These authors even suggested a beneficial

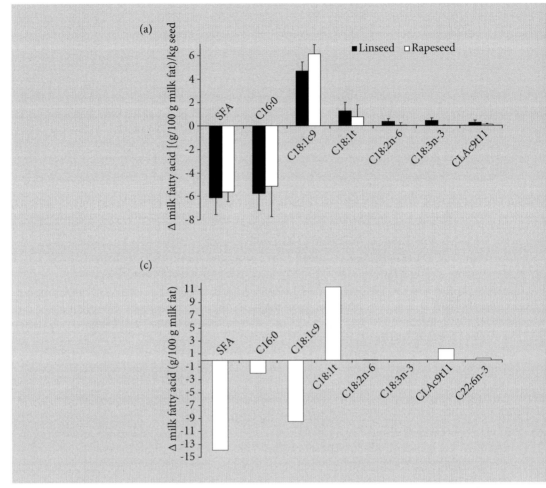

Figure 5. Differences in milk fatty acid composition with supplementation of MCFA and PUFA sources. 1 kg linseed (LIN) (Collomb et al., 2004; Gonthier et al., 2005; Mustafa et al., 2003; Offer et al., 2001) (a and b); rapeseed (Givens et al., 2003; Collomb et al., 2004) (a); soybean

effect of C14:0, but this was based on a limited number of observations. On the other hand, the increase in trans fatty acids upon dietary PUFA supplementation is still under debate with some suggesting a possible positive bioactive role of the main ruminant trans isomer (vaccenic acid, C18:1 trans-11) (e.g. Lock et al., 2008), whereas others suggest similar effects to those of industrial trans fatty acids (e.g. Nestel, 2008). Generally, the trans FA amounts eaten from dairy products probably will be too small to be of concern, at least in dairy products of cows which were not fed marine products.

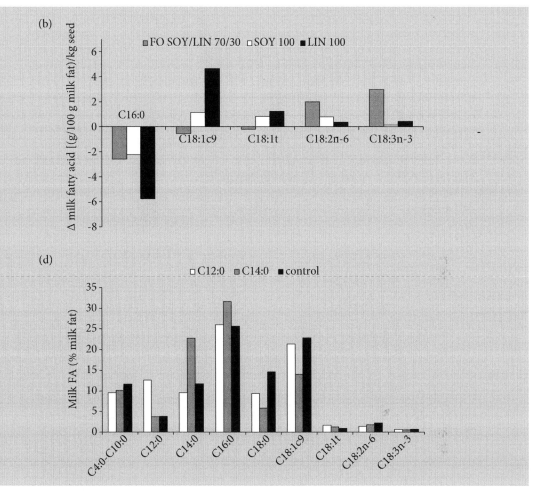

(SOY) (AbuGhazaleh et al., 2002) (b) or formaldehyde treated (FO) soybean/linseed mixture (Gulati et al., 2002) (b) or 200 g marine algae (Boeckaert et al., 2008) (c) or 40 g/kg lauric (C12:0) or myristic (C14:0) acid (Dohme et al., 2004) (d).

8. Oil supplements as methane mitigation strategies: perspectives

MCFA or PUFA supplementation provokes some pressure on the rumen system which results in a risk for reduced feed intake, lower digestibility and hence reduced performance. Moreover, as mentioned in Section 5, FA exert more pronounced effects on methane reduction when supplemented to concentrate based diets. However, increased levels of concentrates also imply higher risks for other rumen health problems e.g. acidosis. Hence, supplementation of MCFA or PUFA rich sources requires 'user-specific' expertise to identify the optimal feeding level, which – to some extent – might differ from farm to farm, e.g. in relation to differences in basal diet. Inversely, this interaction between the basal diet and supplementation level, will lead to variation in the extent of methane reduction despite similar oil supplementation levels. This means individual farm monitoring is required to allow identification of the optimal supplementation level in terms of production parameters as well as greenhouse gas budget. With respect to the latter, governmental incentives to stimulate the application of methane mitigation strategies on a farm level only will be developed when the effectiveness of these strategies can be assessed reliably at the farm level.

Recently, successful efforts have been made by our group to predict rumen proportions of SCFA based on milk odd and branched-chain fatty acids (e.g. Vlaeminck *et al.*, 2006; Craninx *et al.*, 2008). Given the link between rumen methane and SCFA production, milk OBCFA also might show scope to quantify variation in rumen methanogenesis. A first, promising, assessment has been made, based on ruminal methanogenesis, calculated from molar proportions of acetate, propionate and butyrate and assuming a hydrogen recovery of 0.9 as $CH_4 = 0.45 \times Ac - 0.275 \times Pr + 0.4 \times But$. Predictions based on milk OBCFA were suggested to show potential in predicting ruminal methanogenesis but they should be confirmed using direct *in vivo* measurements of rumen CH_4 production (Vlaeminck and Fievez, 2005). This research is currently ongoing at our laboratory.

High costs associated with dietary oil supplementation and a risk to negatively affect milk yield and milk fat concentration (Zheng *et al.* 2005) are the major constraints in using fat supplements in animal diets. Indeed, several have argued that the use of these high cost ingredients was not justified when their supplementation did not increase the metabolisable energy content of the diet. However, mutual benefits from a nutritional as well as environmental point of view might justify higher costs and adequate supplementation and monitoring should avoid negative side effects. Moreover, additional benefits in terms of animal health and fertility have been suggested and research is currently ongoing in this respect.

In summary, supplementing lipids to animal diets increases the energy density of feed and possess potential as a methane mitigating strategy due to antimicrobial activities of individual FA. However, applicability of fat supplements might be hampered by the associative adverse effects on feed intake and milk yields. Therefore, caution should be taken in selecting the appropriate fat source and level of supplementation to have concomitant methane reduction and more milk per unit of methane. Supplementation with MCFA rich sources and PUFA rich oil seeds should be based on their environmental and nutritional value and cost. It is evident both from *in vitro* and *in vivo* literature that MCFA and PUFA possess potential to reduce methane production, However, direct extrapolation of *in vitro* results to *in vivo* situations is difficult (Flachowsky and Lebzien, 2009) and long term studies also should be considered (Woodward *et al.*, 2006). Realistic methane reduction through oil supplements are in the order of magnitude of 10 to 15%.

References

AbuGhazaleh, A.A., Schingoethe, D.J., Hippen, A.R., Kalscheur, K.F., Whitlock, L.A., 2002. Fatty acid profiles of milk and rumen digesta from cows fed fish oil, extruded soybeans or their blend. Journal of Dairy Science 85: 2266-2276.

Anonymous, 2009. http://www.chempro.in/fattyacid.htm.

Baldwin, R.L., Allison, M.J., 1983. Rumen metabolism. Journal of Animal Science 57 (Suppl. 2): 461-477.

Beauchemin, K.A., McGinn, S.M., 2005. Methane emissions from feedlot cattle fed barley or corn diets. Journal of Animal Science 83: 653-661.

Beauchemin, K.A., McGinn, S.M., Benchaar, C., Holtshausen, L., 2009. Crushed sunflower, flax, or canola seeds in lactating dairy cow diets: Effects on methane production, rumen fermentation, and milk production. Journal of Dairy Science 92: 2118-2127.

Boeckaert, C., Arvidsson, K., Boon, N., Fievez, V. 2007. Effect of vitamin E or vitamin C on *in vitro* biohydrogenation of linolenic and linoleic acid in the presence of unesterified DHA. Journal of Animal Science, 85: Suppl. 1, 119.

Boeckaert, C., Mestdagh, J., Vlaeminck, B., Clayton, D., Fievez, V. 2006. Micro-algae as potent rumen methane inhibitors and modifiers of rumen lipolysis and biohydrogenation of linoleic and linolenic acid. International Congress Series 1293: 184-188.

Boeckaert, C., Vlaeminck, B., Dijkstra, J., Issa-Zacharia, A., Van Nespen, T., Van Straalen, W., Fievez, V. 2008. Effect of Dietary Starch or Micro Algae Supplementation on Rumen Fermentation and Milk Fatty Acid Composition of Dairy Cows. Journal of Dairy Science 91: 4714-4727.

Carro, M. D., Ranilla, M.J., 2003. Effect of the addition of malate on in vitro rumen fermentation of cereal grains. British Journal of Nutrition 89: 279-288.

Chilliard, Y., Ferlay, A., 2004. Dietary lipids and forages interactions on cow and goat milk fatty acid composition and sensory properties. Reproduction Nutrition and Development 44: 467-492.

Collomb, M., Sollberger, H., Butikofer, U., Sieber, R., Stoll, W., Schaeren, W., 2004. Impact of a basal diet of hay and fodder beet supplemented with rapeseed, linseed and sunflower seed on the fatty acid composition of milk fat, International Dairy Journal 14: 549-559.

Craninx, M., Fievez, V., Vlaeminck, B., De Baets, B. 2008. Artificial neural network models of the rumen fermentation pattern in dairy cattle. Computers and Electronics in Agriculture 60: 226-238.

Czerkawski J.W., Blaxter K.L., Wainman F.W., 1966a. The effect of functional groups other than carboxyl on the metabolism of C_{18} and C_{12} alkyl compounds by sheep. British Journal of Nutrition 20: 495-508.

Czerkawski J.W., Blaxter K.L., Wainman F.W., 1966b. The metabolism of oleic, linoleic and linolenic acids by sheep with reference to their effects on methane production. British Journal of Nutrition 20: 349-362.

Czerkawski, J.W., Christie, W.W., Breckenridge, G., Hunter, M.L, 1975. Changes in the rumen metabolism of sheep given increasing amounts of linseed oil in their diets. British Journal of Nutrition 34: 25.

Dang Van, Q.C., Focant, M., Deswysen, D., Mignolet, E., Turu, C., Pottier, J., Froidmont, E., Larondelle, Y. 2008. Influence of an increase in diet structure on milk conjugated linoleic acid content of cows fed extruded linseed. Animal 2: 1538-1547.

De Smet, S., 1993. Effects of defaunation on site, nature and extent of digestion in sheep. Thesis submitted to obtain the degree of Doctor (Ph.D.) in Agricultural Sciences. Ghent University, Ghent.

Demeyer, D., Henderickx, H., 1967. The effect of C18 unsaturated fatty acids on methane production in vitro by mixed rumen bacteria. Biochimica Biophysica Acta 37: 484-497.

Demeyer, D.I., Van Nevel, C.J., 1975. Methanogenesis, an integrated part of carbohydrate fermentation, and its control. In: McDonald, I.W., Warner, A.C.I. (eds.) Digestion and Metabolism in the Ruminant. Proceedings of the 4th International Symposium on Ruminant Physiology. University of New England Publishing Unit, Armidale, Australia, pp. 366-382.

Demeyer, D.I., 1991. Quantitative aspects of microbial metabolism in the rumen and hindgut. In: Jouany J.-P. (ed.) Rumen microbial metabolism and ruminant digestion. INRA Editions, Paris, France, pp. 217-237.

Demeyer, D.I., Fievez, V., 2000. Ruminants and environment: methanogenesis. Annales De Zootechnie 49: 95-112.

Dohme F., Machmüller, A.,Sutter, F., Kreuzer M., 2004. Digestive and metabolic utilization of lauric, myristic and stearic acid in cows, and associated effects on milk fat quality. Archives of animal nutrition 58: 99-116.

Dohme, F., Machmüller, A., Wasserfallen, A., Kreuzer, M., 2001. Ruminal methanogenesis as influenced by individual fatty acids supplemented to complete ruminant diets. Letters in Applied Microbiology 32: 47-51.

Doreau, M., Laverroux, S., Normand, J., Chesneau, G., Glasser, F. 2009. Effect of Linseed Fed as Rolled Seeds, Extruded Seeds or Oil on Fatty Acid Rumen Metabolism and Intestinal Digestibility in Cows. Lipids 44: 53-62.

Eugène, M., Massé, D., Chiquette, J., Benchaar, C., 2008. Meta-analysis on the effects of lipid supplementation on methane production in lactating dairy cows. Canadian Journal of Animal Science 88: 331-334.

Fievez, V., Boeckaert, C., Vlaeminck, B., Mestdagh, J., Demeyer D., 2007. In vitro examination of DHA-edible micro algae: 2. Effect on rumen methane production and apparent degradability of hay. Animal Feed Science and Technology 136: 80-95.

Fievez, V., Dohme, F., Danneels, M., Raes, K., Demeyer, D., 2003. Fish oils as potent rumen methane inhibitors and associated effects on rumen fermentation in vitro and in vivo. Animal Feed Science and Technology 104: 41-58.

Flachowsky, G., Lebzien, P., 2009. Comments on in vitro studies with methane inhibitors. Animal Feed Science and Technology 151: 337-339.

Fonty, G., Williams, A.G., Bonnemoy, F., Morvan, B., Withers, S.E., Gouet, P., 1997. Effect of *Methanobreviabacter* sp. MF1 inoculation on glycoside hydrolase and polysaccharide depolymerase activities, wheat straw degradation and volatile fatty acid concentrations in the rumen of gnotobiotically-reared lambs. Anaerobe 3: 383-389.

Freese, E., Sheu, C.W., Galliers, E., 1973. Function of lipophilic acids as antimicrobial food additives. Nature 241: 321-325.

Galbraith, H., Miller, T.B., Paton, A., Thompson, J.K., 1971. Antibacterial activity of long chain fatty acids and the reversal with calcium, magnesium, ergocalciferol and cholesterol. Journal of Applied Bacteriology 34: 803-813.

Giger-Reverdin, S., Morand-Fehr, P., Tran, G., 2003. Literature survey of dietary fat composition on methane production in dairy cattle. Livestock Production Science 82: 73-79.

Givens, D.I., Allison, R., Blake, J.S., 2003. Enhancement of oleic acid and vitamin E concentrations of bovine milk using dietary supplements of whole rapeseed and vitamin E. Animal Research, 52: 531-542.

Goel, G., Arvidsson, K.,Vlaeminck, B., Bruggeman, G., Deschepper, K., Fievez, V., 2009. Effects of capric acid on rumen methanogenesis and biohydrogenation of linoleic and a-linolenic acid. Animal. 3: 810-816.

Gonthier, C., Mustafa, A. F., Ouellet, D.R., Chouinard, P.Y., Berthiaume, R., Petit, H.V., 2005. Feeding micronized and extruded flaxseed to dairy cows: effects on blood parameters and milk fatty acid composition. Journal of Dairy Science 88: 748-756.

Gulati, S.K., May, C., Wynn, P.C., Scott, T.W., 2002. Milk fat enriched in n-3 fatty acids. Animal Feed Science and Technology 98: 143-152.

Hassim, H., Lourenco, M., Goel, G., Goh, Y., Fievez, V., 2008. Effects of different inclusion levels of oil palm fronds on *in vitro* short chain fatty acid and methane productions, and on rumen biohydrogenation. Communications in Agricultural and Applied Biological Sciences 73: 153-156.

Henderson, C., 1973. The effect of fatty acids on pure cultures of rumen bacteria. Journal of Agricultural Science 81: 107-112.

Hino T., Andoh N., Ohgi H., 1993. Effects of beta-carotene and alfa-tocopherol on rumen bacteria in the utilization of long-chain fatty acids and cellulose. Journal of Dairy Science 76: 600-605.

Holovska K., Lenartova V., Holovska K., Pristas P., Javorsky P., 2002. Are ruminal bacteria protected against environmental stress by plant antioxidants? Letters in Applied Microbiology 35: 301-304.

Jenkins, T.C., Palmquist, D.L., 1984. Effect of fatty acids or calcium soaps on rumen and total nutrient digestibility of dairy rations. Journal of Dairy Science 67: 978-986.

Jenkins, T.C., Palmquist, D.L., 1982. Effect of added fat and calcium on *in vitro* formation of insoluble fatty acid soaps and cell wall digestibilitiy. Journal of Animal Science 55: 957-964.

Jordan, E., Kenny, D., Hawkins, M., Malone, R., Lovett, D.K., O'Mara, F. P., 2006a. Effect of refined soy oil or whole soybeans on intake, methane output, and performance of young bulls. Journal of Animal Science 84: 2418-2425.

Jordan, E., Lovett, D.K., Monahan, F.J., Callan, J., Flynn, B., O'Mara, F.P., 2006b. Effect of refined coconut oil or copra meal on methane output and on intake and performance of beef heifers. Journal of Animal Science 84: 162-170.

Keltjens, J.T., Vogels, G.D., 1996. Metabolic regulation in methanogenic archaea during growth on hydrogen and CO_2. Environmental Monitoring and Assessment 42: 19-37.

Keweloh, H., Heipieper, H.J., 1996. Trans unsaturated fatty acids in bacteria. Lipids, 31: 129-137.

Klevenhusen, F., Bernasconi, S.M., Hofstetter, T.B., Bolotin, J., Kunz, C., Soliva, C.R., 2009. Efficiency of monolaurin in mitigating ruminal methanogenesis and modifying C-isotope fractionation when incubating diets composed of either C3 or C4 plants in a rumen simulation technique (Rusitec) system. British Journal of Nutrition (doi: 10.1017/S0007114509990262).

Knapp, H.R., Melly, M.A., 1986. Bactericidal effects of polyunsaturated fatty acids. Journal of Infectious Diseases 154: 84-94.

Lock, A.L., Destaillats, F., Kraft, J., German, J.B., 2008. Introduction to the proceedings of the symposium Scientific update on dairy fats and cardiovascular disease. Journal of the America, College of Nutrition 27: 720S-722S.

Lovett, D.K., Lovell, S., Stack, L., Callan, J., Finlay, M., Conolly, J., O'Mara, F.P., 2003. Effect of forage/concentrate ratio and dietary coconut oil level on methane output and performance of finishing beef heifers. Livestock Production Science 84: 135-146.

Machmüller, A., Kreuzer, M., 1999. Methane suppression by coconut oil and associated effects on nutrient and energy balance in sheep. Canadian Journal of Animal Science 79: 65-72.

Machmüller, A., 2006. Medium-chain fatty acids and their potential to reduce methanogenesis in domestic ruminants. Agriculture, Ecosystems & Environment 112: 107-114.

Machmüller, A., Dohme, F., Soliva, C.R., Wanner, M., Kreuzer, M., 2001. Diet composition affects the level of ruminal methane suppression by medium-chain fatty acids. Australian Journal of Agricultural Research 52: 713-722.

Machmüller, A., Ossowski, D.A., Wanner, M., Kreuzer, M., 1998. Potential of various fatty feeds to reduce methane release from rumen fermentation *in vitro* (Rusitec). Animal Feed Science and Technology 71: 117-130.

Machmüller, A., Soliva, C.R., Kreuzer, M., 2002. *In vitro* ruminal methane suppression by lauric acid as influenced by dietary calcium. Canadian Journal of Animal Science 82: 233-239.

Machmüller, A., Soliva, C.R., Kreuzer, M., 2003a. Methane-suppressing effect of myristic acid in sheep as affected by dietary calcium and forage proportion. British Journal of Nutrition 90: 529-540.

Machmüller, A., Soliva, C.R., Kreuzer, M., 2003b. Effect of coconut oil and defaunation treatment on methanogenesis in sheep. Reproduction Nutrition and Development 43: 41-55.

Maczulak, A.E., Dehority, B.A., Palmquist, D.L., 1981. Effects of long-chain fatty acids on growth of rumen bacteria. Applied and Environmental Microbiology 42: 856-862.

Martin, C., Rouel, J., Jouany, J.P., Doreau, M., Chilliard, Y., 2008. Methane output and diet digestibility in response to feeding dairy cows crude linseed, extruded linseed, or linseed oil. Journal of Animal Science 86: 2642-2650.

McCord, J.M., Keele, B.B., Fridovic, I., 1971. An enzyme-based theory of obligate anaerobiosis: the physiological function of superoxide dismutase. Proceedings of Natural Academy of Science U. S. A. 68: 1024-1027.

Mensink, R.P., Zock, P. L., Kester, A.D., Katan, M.B., 2003. Effects of dietary fatty acids and carbohydrates on the ratio of serum total to HDL cholesterol and on serum lipids and apolipoproteins: a meta-analysis of 60 controlled trials. American Journal of Clinical Nutrition 77: 1146-1155.

Mustafa, A. F., Gonthier, C., Ouellet, D.R., 2003. Effects of extrusion of flaxseed on ruminal and postruminal nutrient digestibilities. Archives of Animal Nutrition 57: 455-463.

Nair, M.K., Joy, J., Vasudevan, P., Hinckley, L., Hoagland, T.A., Venkitanarayanan, K.S., 2005. Antibacterial effect of caprylic acid and monocaprylin on major bacterial mastitis pathogens. Journal of Dairy Science 88: 3488-3495.

Nestel, P.J., 2008. Effects of dairy fats within different foods on plasma lipids. Journal of the American College of Nutrition 27: 735S-740S.

Newbold, C.J., Lassalas, B., Jouany, J.P., 1995. The importance of methanogens associated with ciliate protozoa in ruminal methane production *in vitro*. Letters in Applied Microbiology 21: 230-234.

Newbold, C.J., Chamberlain, D.G., 1988. Lipids as rumen defaunating agents. Proceedings of Nutrition Society 47: 154A. (abstr.).

Nieman, C., 1954. Influence of trace amounts of fatty acids on the growth of microorganisms. Bacteriological Reviews 18.

Odongo, N.E., Or-Rashid, M.M., Kebreab, E., France, J., McBride, B.W., 2007. Effect of supplementing myristic acid in dairy cow rations on ruminal methanogenesis and fatty acid profile in milk. Journal of Dairy Science 90: 1851-1858.

Offer, N.W., Marsden, M., Phipps, R.H., 2001. Effect of oil supplementation of a diet containing a high concentration of starch on levels of trans fatty acids and conjugated linoleic acids in bovine milk. Animal Science 73: 533-540.

Panayakaew, P., Goel, G., Lourenco, M., Yuangklang, C., Fievez, V., 2008. Medium-chain fatty acuds from coconut oil or krabok oil to reduce *in vitro* rumen methanogenesis. Communications in Agricultural and Applied Biological Sciences 73, 189-192.

Pereira, M.A., Pires, O.C., Mota, M., Alvest, M.M., 2005. Anaerobic biodegradation of oleic and palmitic acids: evidence of mass transfer limitations caused by long chain fatty acid accumulation onto the anaerobic sludge. Biotechnology and Bioengineering 92: 15-23.

Prins, R.A., Van Nevel, C.J., Demeyer, D.I., 1972. Pure culture studies of inhibitors for methanogenic bateria. Antonie Van Leeuwenhoek 38: 281-287.

Rohr, R., Daenicke, Oslage, H.J., 1978. Untersuchungen über den Einfluß verschiedener Fettbeimischungen zum Futter auf Stoffwechsel und Leistung von Milchkühen, Landbauforschung Völkenrode 28: 139-150.

Soliva, C.R., Hindrichsen, I.K., Meile, L., Kreuzer, M., Machmüller, A., 2003. Effects of mixtures of lauric and myristic acid on rumen methanogens and methanogenesis *in vitro*. Letters in Applied Microbiology 37: 35- 39.

Soliva, C.R., Meile, L., Cieślak,, A., Kreuzer, M., Machmüller, A., 2004. Rumen simulation technique study on the interactions of dietary lauric and myristic acid supplementation in suppressing ruminal methanogenesis. British Journal of Nutrition 92: 689-700.

Steele, W., Moore, J.H., 1968. The digestibility coefficients of myristic, palmitic and stearic acids in the diet of sheep Journal of Dairy Research 35: 371-376.

Stillwell, W., Wassall, S.R., 2003. Docosahexaenoic acid: membrane properties of a unique fatty acid. Chemistry and Physics of Lipids 126: 1-27.

Ushida, K., Umeda, M., Kishigami, N., Kojima, Y. 1992. Effect of medium chain fatty chain and long chain fatty acids calcium salts on rumen microorganisms and fibre digestion in sheep. Animal Science and Technology (Japan) 63: 591-597.

Van Nevel, C., 1991. Modification of rumen fermentation by the use of additives. In: Jouany J.-P. (Ed.), Rumen microbial metabolism and ruminant digestion. INRA Editions, Paris, pp. 263-280.

Van Nevel, C.J., Demeyer, D.I., 1995. Feed additives and other interventions for decreasing methane emissions. In: Wallace, R., Chesson, A. (eds.) Biotechnology in animal feeds and animal feeding. VCM, Weinheim, pp. 329-349.

Van Soest, P.J., 1982. Nutritional ecology of the ruminant. O & B Books Inc, Corvallis.

Vlaeminck, B., Fievez, V. 2005. Potential of milk odd and branched-chain fatty acids to predict ruminal methanogenesis in dairy cows. In: Soliva, C.R., Takahashi, J., Kreuzer, M. (eds.) Publication Series, Institute of Animal Science, ETH Zurich, 27, pp. 393-396.

Vlaeminck, B., Fievez, V., Tamminga, S., Dewhurst, R.J., Van Vuuren, A., De Brabander, D., Demeyer, D. 2006. Milk Odd- and Branched-Chain Fatty Acids in Relation to the Rumen Fermentation Pattern. Journal of Dairy Science 89: 3954-3964.

Walichnowski, Z., Lawrence, S.G., 1982 Studies into the effects of cadmium and low pH upon methane production. Hydrobiologia 91-92: 1573-5117.

Williams, A.G., Coleman, G.S., 1992. The rumen protozoa. Springer-Verlag, New York.

Wolin, M.J., Miller, T.L., 1988. Microbe – microbe interactions. In: Hobson, P.M. (Ed.), The rumen microbial system. Elsevier Applied Science, London, pp. 343-359.

Woodward S.L., Waghorn G.C., Thompson N.A., 2006. Supplementing dairy cows with oils to improve performance and reduce methane – does it work? Proceedings of NewZealand Soceity of Animal Production 66,176-181.

Zhang, C.M., Guo, Y.Q., Yuan, Z.P., Wu, Y.M., Wang, J.K., Liu, J.X., Zhu, W.Y., 2008. Effect of octadeca carbon fatty acids on microbial fermentation, methanogenesis and microbial flora in vitro. Animal Feed Science and Technology 146: 259-269.

Zheng, H.C., Liu, J.X., Yao, J.H., Yuan, Q., Ye, H.W., Ye, J.A., Wu, Y.M., 2005. Effects of dietary sources of vegetable oils on performance of high-yielding lactating cows and conjugated linoleic acid in milk. Journal of Dairy Science 88: 2037-204.

Nutritional manipulation of subacute ruminal acidosis in dairy cattle

Garret R. Oetzel
School of Veterinary Medicine, University of Wisconsin-Madison, 2015 Linden Drive, Madison, WI 53706, USA; groetzel@wisc.edu

Abstract

Subacute ruminal acidosis (SARA) is an important nutritional disease in dairy cattle. It both impairs animal welfare and causes economic losses. No single test or clinical sign is diagnostic of SARA in a dairy herd. However, combinations of low ruminal pH, high prevalence of lameness, high herd turnover rate (with even distribution across the lactation cycle), occasional cases of nosebleeds or coughing blood, altered faecal consistency, low faecal pH, and/or milk fat depression considered together are indicative of SARA. Key factors in the nutritional management of SARA include limiting the intake of rapidly fermentable carbohydrates, provided adequate physically effective neutral detergent fibre, providing the highest possible proportion of forage in the diet, delivering feed with excellent accuracy and consistency, appropriately processing forages and concentrate feed ingredients, adding long forage particles to high corn silage diets, providing sufficient feed bunk space, providing enough buffering capacity in the diet, providing enough long particles to stimulate chewing, preventing sorting of total mixed rations, and adequately adapting the rumen to higher concentrate diets. Free-choice low-moisture molasses blocks with added buffer and the inclusion of certain direct-fed microbials may provide some additional protection against SARA in dairy herds.

1. Introduction

Ruminal acidosis is the consequence of feeding high grain diets to ruminant animals, which are adapted to digest and metabolize predominantly forage diets. Feeding diets that are progressively higher in grain tends to increase milk production, even in diets containing up to 75% concentrates. However, short-term gains in milk production are often substantially or completely negated by decreased milk fat percentage and long-term compromises in cow health when high grain diets are fed.

Compromises in dairy cow health due to ruminal acidosis are a concern for reasons of good animal welfare as well as for economic reasons. Lameness is probably most important animal welfare issue today in dairy herds, and a good portion of the lameness observed in dairy cows may be attributed to laminitis secondary to high grain feeding (Nordlund *et al.*, 2004). Lameness (along with secondary reproductive failure and low milk production) is commonly the most important cause of premature, involuntary culling and unexplained cow deaths in a dairy herd.

Acute ruminal acidosis (uncompensated decline in ruminal pH, accumulation of ruminal lactate, and obvious clinical signs in affected cows) is uncommon in dairy cattle. Subacute ruminal acidosis (SARA) is more common in dairy herds and is characterized by spontaneous recovery from periods of low ruminal pH, transient or no accumulation of ruminal lactate, and subtle clinical signs during the time of low ruminal pH.

2. Monitoring SARA in dairy herds

A major limitation in our understanding of SARA is the difficulty in monitoring it in dairy herds. There is no single herd-level monitor for SARA; rather, it is evaluated by considering several different factors that are individually limited but can be considered together to monitor SARA.

Subacute ruminal acidosis is defined by some degree of transient depression in ruminal pH is the definition of SARA. However, ruminal pH varies considerably by time after feeding and between individual cows (Krause *et al.*, 2006). Thus, continuous monitoring of a substantial number of cows is necessary to evaluate ruminal pH within a herd. There no practical means to accomplish this with current technology. Spot checks of ruminal pH can be done via rumenocentesis; however, this procedure is limited and can only detect herds with very high prevalences of low ruminal pH (Garrett *et al.*, 1999)

Even using continuous monitoring of ruminal pH, the exact ruminal pH cut point that defines SARA has not been determined. Ruminal pH cut points of 5.5, 5.6, and 5.8 have been proposed. Additionally, the degree to which ruminal pH must be below a certain cut point before it is considered SARA also varies. Examples from the literature include ruminal pH below 5.8 for more than about 5 hours per day (Zebeli *et al.*, 2008) or ruminal pH below 5.6 for more than about 3 hours per day (Plaizier *et al.*, 2009). Criteria for defining SARA by ruminal pH must also take into consideration the location of the sample within the rumen and the method by which it was collected.

A high prevalence of lameness in a herd may indicate SARA, especially if claw horn disorders are the main cause of lameness (Nordlund *et al.*,

2004). However, other factors such as exposure to concrete, type of stall surface, type of walking surface, feed bunk space and design, pen layout, overstocking, heat stress, and stall use behaviour also contribute to claw horn lesions (Cook *et al.*, 2004).

Herds with a high prevalence of SARA may have low body condition scores despite apparently adequate energy density in the diets (Nordlund *et al.*, 1995). This may be explained by secondary complications of SARA that reduce feed intake (e.g., lameness, hepatic abscesses, or renal abscesses) or by the energetic cost of inflammatory responses to rumenitis and other complications of SARA. However, other nutritional problems and disease processes can also cause low body condition scores despite adequate dietary energy density.

Herds with a high prevalence of SARA typically have high herd turnover rates, with removals scattered across the lactation cycle. Other causes of high herd turnover rates are often biased toward early or late lactation. However, other herd problems (or combinations of problems) can cause herd removal patterns similar to those observed with SARA.

Occasional cases of bilateral epistaxis (nosebleed) and/or hemoptysis (coughing up blood) are sometimes observed in herds with a high prevalence of SARA. These signs are almost always caused by caudal vena cava syndrome, which is the showering of the lungs with septic emboli from liver abscesses caused by SARA. These septic emboli eventually cause pulmonary bleeding. However, many herds with substantial SARA have no history of cows showing epistaxis or hemoptysis.

Faecal consistency is inconsistently affected by SARA. Affected herds may have a high prevalence of cows with loose manure that may be bright, yellowish, foamy, have a sweet-sour smell, or contain undigested feed particles. However, changes in faecal characteristics are not a consistent feature of SARA (Kleen *et al.*, 2009). Factors other than SARA, such as protein feeding and hindgut carbohydrate fermentation, also affect faecal consistency. Infectious diseases such as Johne's disease, salmonellosis, or winter dysentery can also affect faecal consistency.

The pH of manure may have some association with SARA. If collected at the same time as ruminal pH, faecal pH is not associated with ruminal pH (Enemark *et al.*, 2004). Limited research data from a SARA challenge model suggest that faecal pH can be associated with ruminal pH, but lags by 6 to 8 hours (Oetzel, unpublished data). It is not known if this relationship exists in typical feeding situations. If the nadir in ruminal pH is reached about 8 to 14 hours after the first feeding of the day (Krause *et al.*, 2005), then the faecal pH nadir would be expected to occur about 14 to 22 hours after this feeding (typically the middle of the night). This is a difficult time to gather and analyze samples.

Milk fat percentage is sometimes depressed by SARA. However, this effect is inconsistent (Enemark, 2009) and may be present only if SARA persists more than a few days (Krause *et al.*, 2005). Milk fat percentage in early lactation cows is particularly unresponsive to SARA (Enemark *et al.*, 2004). Many dietary factors unrelated to SARA also influence milk fat percentage (Bauman *et al.*, 2003).

The challenge to the nutritionist, veterinarian, and dairy producer is to evaluate herd performance in all of these areas and then make a clinical judgment about the presence of absence of SARA based on the combined evidence. No single herd indicator is diagnostic for SARA, but a combination of several factors leading to the same conclusion is sufficient evidence.

3. Nutritional causes of SARA

Excessive intake of rapidly fermentable carbohydrates

Excessive intake of rapidly fermentable carbohydrates is the most obvious and direct cause of SARA in dairy cattle. Total intake of rapidly fermentable carbohydrates is inversely proportional to the intake of fibrous carbohydrates (defined as carbohydrates that are slowly or not fermented in the rumen). Thus, the ability of diets to cause SARA may be expressed either as a measure of fermentable or of fibrous carbohydrates.

Nutritionists have devised a variety of systems to determine the effect of diet on SARA. These include neutral detergent fibre (NDF), forage NDF, non-fibre carbohydrates, starch, and physically effective NDF (peNDF). The peNDF system was devised to predict the fraction of dietary fibre that stimulates chewing and contributes to the mat layer in the rumen. Physically effective NDF is determined by multiplying the NDF content of a dried feed ingredient (or TMR sample) by the proportion of it that is retained on top of a 1.18 mm screen after vertical shaking (Mertens, 1997). Although not designed to predict ruminal pH, peNDF was the dietary measure that most accurately predicted ruminal pH in a recent meta-analysis of previously published data (Zebeli *et al.*, 2008). This study also reported that a peNDF level of about 30 to 33% was optimal for minimizing the risk of SARA and maximizing milk production efficiency. Dry matter intake was slightly depressed at these high concentrations of peNDF; however, milk production efficiency was maximized. High total dry matter intakes were apparently offset by a higher risk for SARA and reduced percentage of milk components. The optimal amount of peNDF suggested by this study is quite high compared to Mertens' original suggestion of a minimum of 21% peNDF.

Determination of peNDF as described by Mertens is primarily limited to research studies because of the practical difficulties in dry sieving feed ingredient and TMR samples. The manually-operated Penn State Particle Separator (PSPS) can be used for on-farm determinations of forage particle length without prior drying of the sample (Kononoff *et al.*, 2003). It is important to note that this technique (which uses undried samples and employs horizontal shaking) is not the same as the research technique (dried samples and vertical shaking) for determining the factor used to calculate peNDF. Even though both methods use the same 1.18 mm screen, the results are considerably different (Yang *et al.*, 2006). Some nutritionists have used results from the PSPS (either above the 1.18 mm screen or above the 8 mm screen) as alternative factors for calculating peNDF. While such approaches may have merit, they have created considerable confusion regarding the definition and application of peNDF.

The relationship between ruminal pH and diet is very complex and multifaceted. No single measure of fibre adequacy can accurately predict ruminal pH. Although peNDF is the best single measure available, it does not account for rumen fermentability, which has a major effect on ruminal pH (Krause *et al.*, 2002). Unfortunately, there is no laboratory test for rumen fermentability.

The proportion of forage in a diet may be an additional determinant of ruminal pH, independent of its influence on peNDF or other indicators of fibre adequacy. Increasing the proportion of forage in a diet helps prevent SARA via the cumulative effect of increased chewing time, increased meal frequency, and decreased ruminal acid production. Low forage diets are problematic because of their inherently high fermentability. Increasing forage particle length cannot compensate for lack of forage in the diet (Yang *et al.*, 2009).

Limited as it may be, evaluating the dietary content nutrients is an important first step in determining the cause of SARA in a dairy herd. This requires a careful evaluation of the diet actually being consumed by the cows. A cursory evaluation of the 'paper' ration formulated by the herd nutritionist is usually of little value. Ascertaining the diet actually consumed by the cows requires a careful investigation of how feed is delivered to the cows, accurate weights of the feed delivered, and updated nutrient analyses of the feeds delivered (particularly the dry matter content of the fermented feed ingredients). Careful bunk sampling and wet chemistry analyses of total mixed rations (TMR) may uncover unknown errors in feed composition or feed delivery (Krause *et al.*, 2006).

Dairy herds that use component feeding in early lactation often increase grain feeding in early lactation at a more rapid rate then the cow's increase in dry matter intake. This puts cows at great risk for SARA, since they cannot eat enough forage to compensate for the extra grain consumed.

Grains that are finely ground, steam-flaked, extruded, and/or very wet will ferment more rapidly and completely in the rumen than unprocessed or dry grains, even if their chemical composition is identical. Similarly, starch from wheat or barley is more rapidly and completely fermented in the rumen that starch from corn. Corn silage that is very wet, finely chopped, or kernel-processed also poses a greater risk for SARA than drier, coarsely chopped, or unprocessed corn silage.

Particle size analysis of grains is a useful adjunct test when assessing the risk for SARA in a dairy herd. Very finely ground grains, especially if they are moist, will increase their rate of fermentation in the rumen and increase the risk for SARA.

Feeding a large proportion of a lactation diet as corn silage often puts cows at higher risk for SARA compared to diets containing more dry hay or hay crop silages. Corn silages vary considerably in the amount of corn grain that they contain and in the extent of processing of that grain (e.g., kernel processing).

Corn silage is also difficult to feed because it typically does not contribute enough long particles to a TMR. Very long chopping of corn silage is not recommended, because it impairs fermentation and increases the risk for sorting at the feed bunk. It is a common (and necessary) practice to add chopped long-stem dry hay or chopped dry straw to TMR containing a high proportion of the forage as corn silage. However, it can be difficult to process the dry forage so that it distributes evenly throughout the TMR and so that the cows cannot sort it. Vertical mixers or grinding the dry forage before adding it to the mixer is often necessary.

Feed delivery and feed access are often overlooked as risk factors for SARA. Dairy cattle groups are commonly fed for *ad libitum* intake (typically a 5% daily feed refusal) in order to maximize potential dry matter intake and milk yield. However, slightly limiting intake in dairy cattle at high risk for SARA would in theory reduce their risk of periodic over-consumption and SARA. Feed efficiency would likely be improved. This approach has been successfully used in beef feedlots. However, dairy cow groups are much more dynamic than feedlot groups. This makes it considerably more challenging for dairy cattle feeders to slightly limit intakes without letting the feed bunks be without palatable feed more than about four hours a day. It can be done, but only with adequate bunk space and excellent feed bunk management. Perhaps *ad libitum* feeding with a 5% daily feed refusal is the best option for most dairy herds. This would especially apply to the pre- and post-fresh cow groups because they have rapid turnover and because individual cows have rapidly changing dry matter intakes during these time periods.

Meal size is likely an important aspect of nutritional management of SARA. Cows are often able to self-regulate their ruminal pH if they have continuous and predictable access to the same TMR every day. However, even modest bouts of feed restriction can cause cows to subsequently consume meals that are too large. Therefore, good feed bunk management practices are critical SARA prevention – even when chemical fibre, particle length, and grain processing are optimal. Higher forage diets have the added benefit of decreasing meal size and increasing meal frequency (Yang *et al.*, 2009).

Primiparous cows have lower dry matter intakes than older cows; thus, it seems that they should be a lower risk for SARA. However, results of several investigations indicate that primiparous cows are at higher risk for SARA (Enemark *et al.*, 2004; Krause *et al.*, 2006). Primiparous cows may need time to learn to regulate their feed intake when introduced to a high-energy diet for the first time after calving. They may also have difficulty gaining access to feed bunks when older cows are present in the same group. This might lead to larger and less frequent meals, which could increase the risk for SARA. Overstocking of pens alters feeding and social behaviour of dairy cattle by decreasing feeding time and increasing standing time, especially for cows ranked lower in the social hierarchy (Huzzey *et al.*, 2006).

Dairy cattle fed in pasture-based systems are also at risk for SARA (Bramley *et al.*, 2008; O'Grady *et al.*, 2008). Grass pastures may contain high concentrations of rapidly fermentable carbohydrate and may also be low in physically effective fibre. Excessive grain supplementation should be avoided when pasture is the main source of forage. It is likely that dairy cattle with SARA in pasture systems do not development lameness as readily as cattle in conventional housing because cows on pasture have little or no exposure to concrete. However, other adverse effects of SARA (rumenitis, hepatic abscesses, etc.) may similarly affect pastured cattle.

An important goal of effective dairy cow nutrition is to feed as much concentrate as possible, in order to maximize production, without causing ruminal acidosis. This is a difficult and challenging task in the field because the indications of feeding excessive amounts of fermentable carbohydrates (decreased dry matter intake and milk production) are very similar to the results from feeding excessive fibre (again, decreased dry matter intake and milk production). An important distinction is that even slightly over-feeding fermentable carbohydrates causes chronic health problems due to SARA, while slightly under-feeding fermentable carbohydrates does not compromise cow health.

Inadequate ruminal buffering

Ruminant animals have a highly developed system for buffering the organic acids produced by ruminal fermentation of carbohydrates. While the total effect of buffering on ruminal pH is relatively small, it can still account for the margin between health and disease in dairy cows fed large amounts of fermentable carbohydrates. Ruminal buffering has two aspects – dietary and endogenous buffering.

Dietary buffering is the inherent buffering capacity of the diet and is largely explained by dietary cation-anion difference (DCAD). Diets high in Na and K relative to Cl and S have higher DCAD concentrations, tend to support higher ruminal pH, and increase dry matter intake and milk yield. Lactating diets should contain between about +230 to +330 mEq/kg DCAD (Chan *et al.*, 2005). Formulating diets with a high DCAD typically requires the addition of buffers such as sodium bicarbonate or potassium carbonate. Alfalfa forages tend to have a higher DCAD than corn silage, although this depends considerably on the mineral composition of the soil on which they were grown. Concentrate feeds typically have low or negative DCAD, which adds to their already high potential to cause ruminal acidosis because of their high fermentable carbohydrate content.

Endogenous buffers are produced by the cow and secreted into the rumen via the saliva. The amount of physical fibre in the diet determines the extent of buffer production by the salivary glands. Coarse, fibrous feeds contain more effective fibre and stimulate more saliva production during eating than do finely ground feeds or fresh pasture. Coarse, fibrous feeds also make up the mat layer of the rumen, which is the stimulus for rumination.

The ability of a diet and feeding system to promote maximal amounts of ruminal buffering should be evaluated as part of the work-up of a herd diagnosed with SARA. Wet chemistry analysis of a carefully collected TMR bunk sample can be particularly effective in determining the actual DCAD of the diet delivered to the lactation cows. Diets with measured DCAD values below about +230 mEq/kg of (Na + K) – (Cl + S) should be supplemented with additional buffers to provide more Na or K relative to Cl and S.

Endogenous buffering can be estimated by observing the number of cows ruminating (a goal is at least 40% of cows ruminating at any given time) and by measuring the particle length of the TMR actually consumed by the cows using the PSPS. Diets with less than 7% long particles put cows at increased risk for SARA, particularly if these diets are also borderline or low in chemical fibre content (Grant *et al.*, 1990). Increasing chemical fibre content of the diet may compensate for short particle length (Beauchemin *et al.*, 1994).

Diets with excessive (over about 15%) long forage particles can paradoxically increase the risk for SARA. This happens when the long particles are unpalatable and sortable. Sorting of the long particles occurs soon after feed delivery, causing the cows to consume a diet that is low in physically effective fibre after feeding. The diet consumed later in the feeding period is then excessively high in physically effective fibre and low in energy. Socially dominant cows are particularly susceptible to SARA in this scenario, since they are likely to consume more of the fine TMR particles soon after feed delivery. Cows lower on the peck order then consume a very low energy diet. Thus, cows on both ends of the social spectrum become thin and produce poorly. Overstocking cows appears to increase the risk for increased TMR sorting (Hosseinkhani *et al.*, 2008).

It is very difficult to quantitatively evaluate the extent that a TMR is sorted. The most rigorous approach is to gather representative samples of the TMR at approximately 2 hours after feeding and then do particle length analysis at each time point. Gathering representative TMR bunk samples is tedious (gather 12 or more representative along the length of the bunk, mix, and, and then shake down two six-cup subsamples), and repeating this procedure six to ten times over the course of a day is not very practical. A more reasonable approach is to first evaluate the particle length, coarseness of the long forage particles, and dry matter of the TMR. If the proportion of long particles is <15%, if the long particles are not coarse stemmy hay, and the TMR dry matter is below 50%, then it is probably unnecessary to do any further evaluation of TMR sorting. If there are problems in one or more of these areas, then it is practical to start by comparing particle lengths of TMR refusals to the particle lengths of the TMR offered. If the refusals contain no more than about 5 to 10% more total long particles than the TMR offered, then sorting is unlikely to be a major issue. For example, if the TMR offered contains 18% long particles and the TMR refusal is 24% long particles, then sorting is probably not a major issue. But if the TMR refusal contains >28% long particles, then this is cause for concern.

The most common cause of excessive TMR sorting is the inclusion of unprocessed, coarse, dry baled hay in a TMR. Despite the claims of manufacturers, most TMR mixers (except for some vertical mixers) are unable to adequately reduce the particle size of coarse dry hay. Processing this hay before adding it to the mixer is often necessary. In many cases, the dry hay can be eliminated from the TMR, provided there are adequate long particles from haylage and corn silage. The risk for SARA in a herd can sometimes be lowered by removing the baled hay from the TMR.

Provision of free-choice low-moisture molasses blocks containing buffer helped decrease the severity and duration of episodes of low ruminal pH following a SARA challenge (Krause *et al.*, 2009). Free choice buffer intake did not increase during periods of low ruminal pH, suggesting that cows do not have the nutritional wisdom to consume their own buffers as needed. Rather, the benefits of the buffer blocks in this study were apparently due to consistent daily intake.

Inadequate adaptation to highly fermentable, high carbohydrate diets

Cows in early lactation should in theory be particularly susceptible to SARA if they are poorly prepared for the lactation diet they will receive. Ruminal adaptation to diets high in fermentable carbohydrates apparently has two key aspects – microbial adaptation (particularly the lactate-utilizing bacteria, which grow more slowly than the lactate-producing bacteria) and ruminal papillae length (longer papillae promote greater VFA absorption and thus lower ruminal pH) (Dirksen *et al.*, 1985).

The known principles of ruminal adaptation suggest that increasing grain feeding toward the end of the dry period should decrease the risk for SARA in early lactation cows. However, increased grain feeding the dry period had no beneficial effect on early lactation ruminal pH or dry matter intake (Andersen *et al.*, 1999). The practical impacts of ruminal adaptation may be small or even inconsequential in dairy herds – particularly when cows are fed a TMR after calving.

4. Prevention of subacute ruminal acidosis in dairy herds

The basic principles of preventing SARA in dairy herds have been discussed above and include limiting the intake of rapidly fermentable carbohydrates, providing adequate ruminal buffering, and allowing for ruminal adaptation to high grain diets. However, SARA will likely remain an important dairy cow problem even when these principles are understood and applied, because the line between optimal milk production and over-feeding grain is exceedingly fine. In many dairy herd situations, overfeeding grain will transiently increase milk production and see beneficial. However, the long-term health and economic consequences caused by SARA can be devastating. Furthermore, once a cow experiences an episode of SARA, she becomes more prone to further bouts of increasingly severe acidosis (Dohme *et al.*, 2008). Any additional nutritional interventions that might prevent SARA without unduly limiting grain feeding are highly desirable.

Direct-fed microbials

Live yeast (*Saccharomyces cerevisiae*) have been shown to strengthen reducing conditions in the ruminal environment, prevent the accumulation of lactate, and stabilize ruminal pH (Marden *et al.*, 2008). *Aspergillus oryzae* had a modest effect in stabilizing ruminal pH, although interestingly it was beneficial only at the lower dose evaluated (Chiquette, 2009).

A novel application of direct-fed microbials for SARA prevention is to provide lactate producers that provide a steady but small source of lactate in the rumen. In theory this enhances the ruminal lactate-utilizing bacteria and improves ruminal responsiveness to SARA. A mixture of *Enterococcus faecium* (a lactate producer) and *Saccharomyces cerevisiae* provided the best stabilization of ruminal pH in a SARA challenge study (Chiquette, 2009).

Malate supplementation

Selenomonas ruminantium is one of the bacteria that convert ruminal lactate to VFA. *S. ruminantium* is apparently stimulated to utilize lactate by malate (Martin *et al.*, 1995). Supplementing diets with malate as a feed additive may be cost-prohibitive; however, incorporation of forage varieties high in malate may allow for economical inclusion of malate in dairy diets (Callaway *et al.*, 2000).

Supplementation with ionophores

Feeding ionophores reduces ruminal lactate production; this effect appears to be caused by inhibition of lactate-producing bacteria, competitive enhancement of lactate utilizers, and possibly by reducing meal size (Owens *et al.*, 1998). Monensin is approved for use in lactating dairy cattle in the US to improve feed efficiency. However, it does not appear to stabilize ruminal pH in lactating dairy cows, at least during early lactation (Fairfield *et al.*, 2007). Monensin is probably more effective in preventing acute ruminal acidosis, which is characterized by very high ruminal lactate concentrations.

References

Andersen, J.B., Sehested, J., Ingvartsen, L., 1999. Effect of dry cow feeding strategy on rumen pH, concentration of volatile fatty acids and rumen epithelium development. Acta Agriculturae Scandinavica, Section A - Animal Science 49: 149-155.

Bauman, D.E., Griinari, J.M., 2003. Nutritional regulation of milk fat synthesis. Annual Review of Nutrition 23: 203-227.

Beauchemin, K.A., Farr, B.I., Rode, L.M., Schaalje, G.B., 1994. Effects of alfalfa silage chop length and supplementary long hay on chewing and milk production of dairy cows. Journal of Dairy Science 77: 1326-1339.

Bramley, E., Lean, I.J., Fulkerson, W.J., Stevenson, M.A., Rabiee, A.R., Costa, N.D., 2008. The definition of acidosis in dairy herds predominantly fed on pasture and concentrates. Journal of Dairy Science 91: 308-321.

Callaway, T.R., Martin, S.A., Wampler, J.L., Hill, N.S., Hill, G.M., 2000. Malate content of forage varieties commonly fed to cattle. Journal of Dairy Science 80: 1651-1655.

Chan, P.S., West, J.M., Bernard, J.K., Fernandez, J.M., 2005. Effects of dietary cation-anion difference on intake, milk yield, and blood components of the early lactation cow. Journal of Dairy Science 88: 4384-4392.

Chiquette, J., 2009. Evaluation of the protective effect of probiotics fed to dairy cows during a subacute ruminal acidosis challenge. Animal Feed Science and Technology 153: 278-291.

Cook, N.B., Nordlund, K.V., Oetzel, G.R., 2004. Environmental influences on claw horn lesions associated with laminitis and subacute ruminal acidosis in dairy cows. Journal of Dairy Science 87 (E Suppl.): E36-E38.

Dirksen, G.U., Liebich, H.G., Mayer, E., 1985. Adaptive changes of the ruminal mucosa and their functional and clinical significance. The Bovine Practitioner 20: 116-120.

Dohme, F., DeVries, T.J., Beauchemin, K.A., 2008. Repeated ruminal acidosis challenges in lactating dairy cows at high and low risk for developing acidosis: Ruminal pH. Journal of Dairy Science 91: 3554-3567.

Enemark, J.M.D., 2009. The monitoring, prevention and treatment of sub-acute ruminal acidosis (SARA): A review. The Veterinary Journal 176: 32-43.

Enemark, J.M.D., Jorgensen, R.J., Kristensen, N.B., 2004. An evaluation of parameters for the detection of subclinical rumen acidosis in dairy herds. Veterinary Research Communications 28: 687-709.

Fairfield, A.M., Plaizier, J.C., Duffield, T.F., Lindinger, M.I., Bagg, R., Dick, P., McBride, B.W., 2007. Effects of prepartum administration of a monensin controlled release capsule on rumen pH, feed intake, and milk production of transition dairy cows. Journal of Dairy Science 90: 937-945.

Garrett, E.F., Pereira, M.N., Nordlund, K.V., Armentano, L.E., Goodger, W.J., Oetzel, G.R., 1999. Diagnostic methods for the detection of subacute ruminal acidosis in dairy cows. Journal of Dairy Science 82: 1170-1178.

Grant, R.J., Colenbrander, V.F., Mertens, D.R., 1990. Milk fat depression in dairy cows: Role of silage particle size. Journal of Dairy Science 73: 1834-1842.

Hosseinkhani, A., DeVries, T.J., Proudfoot, K.L., Valizadeh, R., Von Keyserlingk, M.A.G., 2008. The effects of feed bunk competition on the feed sorting behavior of close-up dry cows. Journal of Dairy Science 91: 1115-1121.

Huzzey, J.M., DeVries, T.J., Valois, P., Von Keyserlingk, M.A.G., 2006. Stocking density and feed barrier design affect the feeding and social behavior of dairy cattle. Journal of Dairy Science 89: 126-133.

Kleen, J.L., Hooijer, G.A.R.J., Noordhuizen, J.P.T.M., 2009. Subacute ruminal acidosis in Dutch dairy herds. The Veterinary Record 164: 681-684.

Kononoff, P.J., Heinrichs, A.J., Buckmaster, D.R., 2003. Modification of the Penn State forage and total mixed ration particle separator and the effects of moisture content on its measurements. Journal of Dairy Science 86: 1858-1863.

Krause, K.M., Combs, D.K., Beauchemin, K.A., 2002. Effects of forage particle size and grain fermentability in midlactation cows. II. Ruminal pH and chewing activity. Journal of Dairy Science 85: 1947-1957.

Krause, K.M., Dhuyvetter, D.V., Oetzel, G.R., 2009. Effect of a low-moisture buffer block on ruminal pH in lactating dairy cattle induced with subacute ruminal acidosis. Journal of Dairy Science 92: 352-364.

Krause, K.M., Oetzel, G.R., 2005. Inducing subacute ruminal acidosis in lactating dairy cows. Journal of Dairy Science 88: 3633-3639.

Krause, K.M., Oetzel, G.R., 2006. Understanding and preventing subacute ruminal acidosis in dairy herds: A review. Animal Feed Science and Technology 126: 215-236.

Marden, J.P., Julien, C., Monteils, V., Auclair, E., Moncoulon, R., Bayourthe, C., 2008. How does live yeast differ from sodium bicarbonate to stabilize ruminal pH in high-yielding dairy cows? Journal of Dairy Science 91: 3528-3535.

Martin, S.A., Streeter, M.N., 1995. Effect of malate on in vitro mixed ruminal microorganism fermentation. Journal of Animal Science 73: 2141-2145.

Mertens, D.R., 1997. Creating a system for meeting the fiber requirements of dairy cows. Journal of Dairy Science 80: 1463-1481.

Nordlund, K.V., Cook, N.B., Oetzel, G.R., 2004. Investigation strategies for laminitis problem herds. Journal of Dairy Science 87 (E Suppl.): E27-E35.

Nordlund, K.V., Garrett, E.F., Oetzel, G.R., 1995. Herd-based rumenocentesis: A clinical approach to the diagnosis of subacute rumen acidosis. Compendium on Continuing Education for the Practicing Veterinarian 17: S48-S56.

O'Grady, L., Doherty, M.L., Mulligan, F.J., 2008. Subacute ruminal acidosis (SARA) in grazing Irish dairy cows. The Veterinary Journal 176: 44-49.

Owens, F.N., Secrist, D.S., Hill, W.J., Gill, D.R., 1998. Acidosis in cattle: A review. Journal of Animal Science 76: 275-286.

Plaizier, J.C., Krause, D.O., Gozho, G.N., McBride, B.W., 2009. Subacute ruminal acidosis in dairy cows: The physiological causes, incidence and consequences. The Veterinary Journal 176: 21-31.

Yang, W.Z., Beauchemin, K.A., 2006. Physically effective fiber: Method of determination and effects on chewing, ruminal acidosis, and digestion by dairy cows. Journal of Dairy Science 89: 2618-2633.

Yang, W.Z., Beauchemin, K.A., 2009. Increasing physically effective fiber content of dairy cow diets through forage proportion versus forage chop length: Chewing and ruminal pH. Journal of Dairy Science 92: 1603-1615.

Zebeli, Q., Dijkstra, J., Tafaj, M., Steingass, H., Ametaj, B.N., Drochner, W., 2008. Modeling the adequacy of dietary fiber in dairy cows based on the responses of ruminal pH and milk fat production to composition of the diet. Journal of Dairy Science 91: 2046-2066.

Keyword index